Synthesis Lectures on Mathematics & Statistics

Series Editor

Steven G. Krantz, Department of Mathematics, Washington University, Saint Louis, USA

This series includes titles in applied mathematics and statistics for cross-disciplinary STEM professionals, educators, researchers, and students. The series focuses on new and traditional techniques to develop mathematical knowledge and skills, an understanding of core mathematical reasoning, and the ability to utilize data in specific applications.

Hugo D. Junghenn

Symbolic Mathematics with Python

 Springer

Hugo D. Junghenn
George Washington University
Vienna, VA, USA

ISSN 1938-1743   ISSN 1938-1751 (electronic)
Synthesis Lectures on Mathematics & Statistics
ISBN 978-3-031-90521-6   ISBN 978-3-031-90522-3 (eBook)
https://doi.org/10.1007/978-3-031-90522-3

© The Editor(s) (if applicable) and The Author(s), under exclusive license to Springer Nature Switzerland AG 2025

This work is subject to copyright. All rights are solely and exclusively licensed by the Publisher, whether the whole or part of the material is concerned, specifically the rights of translation, reprinting, reuse of illustrations, recitation, broadcasting, reproduction on microfilms or in any other physical way, and transmission or information storage and retrieval, electronic adaptation, computer software, or by similar or dissimilar methodology now known or hereafter developed.
The use of general descriptive names, registered names, trademarks, service marks, etc. in this publication does not imply, even in the absence of a specific statement, that such names are exempt from the relevant protective laws and regulations and therefore free for general use.
The publisher, the authors and the editors are safe to assume that the advice and information in this book are believed to be true and accurate at the date of publication. Neither the publisher nor the authors or the editors give a warranty, expressed or implied, with respect to the material contained herein or for any errors or omissions that may have been made. The publisher remains neutral with regard to jurisdictional claims in published maps and institutional affiliations.

This Springer imprint is published by the registered company Springer Nature Switzerland AG
The registered company address is: Gewerbestrasse 11, 6330 Cham, Switzerland

If disposing of this product, please recycle the paper.

*For my future grandchildren,
may you have happy and fulfilling lives*

Preface

The goal of this book is to give a hands-on approach to computer symbolic computation using elementary commands of Python. The book treats the symbolic manipulation of expressions involving rational functions, logic statements, and the exact fractional solutions of systems of linear equations. The book also contains a variety of applications, including symbolic differentiation and integration. A module with immediately runnable code is available on GitHub for each chapter, allowing the reader to experiment with, modify, or build upon the programs.

Numbers used in computer numerical computations are typically floating point numbers or integers, both of which are restricted in size due to the computer's lack of ability to represent some numbers exactly. This results in rounding errors, which may accumulate during run time, a serious defect in some areas of scientific computing, where exact answers may be needed. Symbolic computation provides exact results, with numbers written as exact fractions and variables processed symbolically. Of course, fractions may eventually need to be approximated by decimal values, but delaying this conversion until the end avoids accumulation of errors and so results in better approximations.

The field of symbolic computation can be quite abstract, delving deeply into the study of algorithms that manipulate mathematical expressions. There are sophisticated software products, for example Mathematica and Maple, that perform complex symbolic computations based on these algorithms. Such commercial products also have extensive graphics capabilities. Additionally, there are computer languages, including Python, with packages that perform some symbolic computations.

How then does this book fit into field of symbolic computation? First, here's what it isn't: It is not a book on abstract symbolic computation algorithms. Nor is it a book on how to use commercial products or computer language packages that implement these algorithms. Indeed, there are excellent books and manuals that fulfil these functions. Rather, the goal of this book is to give a coding approach to symbolic computation using elementary commands of Python. In light of the availability of commercial packages, the reader might reasonably ask, "Why bother?". It's a good question, particularly for those whose interests lie in end results for applications. But for those interested in how some of these results may be achieved, the book may have some relevance. The author is quick note that he does not have any insight into the proprietary code of commercial products. However, many aspects of symbolic computing can be analyzed and then implemented as Python programs in a natural and straightforward way. We have attempted to do this in the book. Of course, the programs developed here do not pretend to compete with professional software packages, these usually being the efforts of teams of mathematicians and programmers and requiring considerable time to develop. Nevertheless, it is hoped that the methods in the book will give the reader some insight into concrete symbolic computer mathematics at an elementary level.

The book is organized as follows. Chapter 1 provides the basic concepts of Python needed to develop the modules in the book. The chapter is anything but encyclopedic. Excellent manuals are available that discuss in detail the multitude of features in Python. The chapter is meant to get the reader off and running with as few coding frills as possible. The module associated with this chapter is Essentials.py.

Chapter 2 collects together many of the common tools required by the programs in the book. These are primarily concerned with scanning mathematical expressions, retrieving various symbols, and inserting symbols into these expressions. The module associated with this chapter is Tools.py.

Chapter 3 develops programs that generate truth tables from expressions involving logical operators and, conversely, programs that generate such expressions from truth tables. The module associated with this chapter is Logic.py.

Chapter 4 develops some elementary number theory, including the division algorithm, the greatest common divisor, prime numbers, congruences and modular arithmetic. One of the main functions here expands a positive integer into a product of primes, giving concrete illustrations of the Fundamental Theorem of Arithmetic. The module associated with this chapter is Number.py.

Chapter 5 constructs a module that simplifies arithmetic expressions involving integers and complex fractions, providing exact results. The main function is the underpinning of subsequent symbolic algebra and calculus programs. A new cipher based on the program is given as an application. The module associated with this chapter is Arithmetic.py.

Chapter 6 develops a module that does one-variable symbolic polynomial algebra. Applications are made to polynomial calculus and interpolation. The module associated with this chapter is PolyAlg.py.

Chapter 7 continues the theme of the preceding chapter, developing programs that extract rational roots and factors from polynomials. The module associated with this chapter is PolyDiv.py.

Chapter 8 generalizes the module in Chap. 6 to rational expressions in several variables. Applications are made to multivariable calculus, including a program that symbolically calculates partial derivatives of multivariable rational functions and one that generates Taylor polynomials in two variables. The module associated with this chapter is MultiAlg.py.

Chapter 9 is the first of a sequence of chapters devoted to symbolic linear algebra. The first half of the chapter develops a program that generates the reduced row echelon form of a matrix with rational complex entries. The second half uses the echelon form to construct exact symbolic solutions of systems of linear equations. The module associated with this chapter is LinSolve.py.

Chapter 10 develops programs that manipulate matrices algebraically, including a matrix calculator that produces symbolic results. An application to curve fitting is given. The module associated with this chapter is MatAlg.py.

Chapter 11 develops programs centering around linear independence of vectors. The main results concern the range and kernel of a matrix. The module associated with this chapter is Vectors.py.

Chapter 12 constructs programs that symbolically evaluate determinants. Applications include a symbolic version of Cramer's rule as well as some geometry. A final application generates the inverse of a matrix with rational function entries. The module associated with this chapter is Determinants.py.

Chapter 13 generalizes the main program in Chap. 8 by allowing expressions that contain not only variables but also parameters. The main function is used to construct a program that outputs the partial fraction decomposition of a rational function. The module associated with this chapter is MultiAlgParams.py.

Much of the mathematics in the book is self-contained, although details are often omitted, the emphasis being on coding. The material is easily accessible to readers with a background in basic calculus and linear algebra.

As noted earlier, all code is available on GitHub: `https://github.com/hjunghenn/Python_Code_for_Computer_Symbolic_Mathematics`.

We should mention that the code here is not necessarily the most concise or efficient. Precedence was given to readability, which frequently entailed introducing extra variables, statements, or functions. The reader may wish to keep this in mind when constructing code. Nothing is more frustrating than writing code that you can't understand two weeks later. (The author admits to having done this more than once.)

Happy coding!

Vienna, VA, USA   Hugo D. Junghenn

Contents

1	**Python Essentials**	1
	1.1 Getting Started	1
	1.2 Functions and Methods	1
	1.3 Values and Variables	2
	1.4 Numerical Operations	4
	1.5 Boolean Operations	6
	1.6 String Operations	7
	1.7 ASCII Functions	10
	1.8 Lists	11
	1.9 Tuples	15
	1.10 Sets	15
	1.11 Dictionaries	16
	1.12 The If Elif Else Statement	16
	1.13 The While Loop	18
	1.14 The For Loop	20
	1.15 Recursion	22
	1.16 Modules	24
2	**General Tools**	25
	2.1 Mathematical Expressions	25
	2.2 Category Functions	26
	2.3 Index Functions	27
	2.4 Retrieving Symbols	28
	2.5 Preparing Expressions	29
	2.6 Extracting Character Groups	32
	2.7 Inserting and Replacing Characters	33
	2.8 List Functions	34
	2.9 Printing Functions	37

3	**Symbolic Logic**		41
	3.1	Compound Statements	41
	3.2	Generating a Truth Table	42
	3.3	The Calculation Engine	46
	3.4	Equivalent Statements	48
	3.5	Valid Arguments	49
	3.6	Disjunctive Normal Form	51
	3.7	Conjunctive Normal Form	52
4	**Properties of Integers**		55
	4.1	Number Bases	55
	4.2	Divisibility	60
	4.3	Extended Euclidean Algorithm	60
	4.4	Multi-extended Euclidean Algorithm	64
	4.5	Least Common Multiple	66
	4.6	The Sieve of Eratosthenes	67
	4.7	The Fundamental Theorem of Arithmetic	68
	4.8	Modular Arithmetic	70
5	**Arithmetic**		75
	5.1	The Main Function	75
	5.2	Conversion Functions	76
	5.3	Arithmetic Operations on Fractions	79
	5.4	Complex Operations	80
	5.5	The Allocator	82
	5.6	Rendering a Fraction into a Decimal	83
	5.7	Converting to Scientific Notation	85
	5.8	Evaluating an Expression	86
	5.9	Ordering Fractions	88
	5.10	Application: Roots by Interval Halving	89
	5.11	A Linear Fractional Cipher	92
6	**Polynomial Algebra**		95
	6.1	Polynomial Operations	95
	6.2	The Main Function	96
	6.3	Polynomial Operations in Python	97
	6.4	The Allocator	99
	6.5	Generating the Polynomial Lists	101
	6.6	Converting Lists to Polynomials	102
	6.7	The Modular Case	104

	6.8	Application: Completing the Square	105
	6.9	Application: Lagrange Interpolation	106
	6.10	Application: Polynomial Calculus	108
	6.11	Application: Special Polynomials	114
7	**Polynomial Divisibility and Roots**	119
	7.1	Division Algorithm for Polynomials	119
	7.2	Extended GCD for Polynomials	121
	7.3	Rational Roots and Linear Factors	123
	7.4	Modular Division Algorithm	125
	7.5	Modular Extended Greatest Common Divisor	127
	7.6	Modular Roots and Factors	128
8	**Multivariable Algebra**	131
	8.1	Rational Functions and Their Representations	131
	8.2	Overview	133
	8.3	Combining Monomials	134
	8.4	Scalar and Variable Conversion Functions	135
	8.5	Calculations	136
	8.6	The Allocator	139
	8.7	Sorting	140
	8.8	Coefficient Reduction	141
	8.9	Variable Reduction	142
	8.10	Clearing Fractions	143
	8.11	Converting a List into a Rational Function	144
	8.12	Rational Function with Integer Coefficients	146
	8.13	Evaluating an Expression	146
	8.14	Application: Partial Differentiation of Rational Functions	147
9	**Linear Equations**	155
	9.1	Matrices	156
	9.2	Systems of Linear Equations	157
	9.3	The Gauss-Jordan Method	157
	9.4	Row Operations in List Form	161
	9.5	Implementing Row Operations	163
	9.6	Row Echelon Form in Python	165
	9.7	Reduced Column Echelon Form	166
	9.8	Linsolve	168
	9.9	Setting Up the Variables	171
	9.10	Creating the Augmented Matrix	172
	9.11	Generating the Solutions	173
	9.12	Checking the Solution	174

10 Matrix Algebra 177
10.1 Elementary Matrix Operations 177
10.2 The Inverse of a Matrix 183
10.3 Matrix Exponentiation 186
10.4 Solving Systems Using Matrix Inversion 187
10.5 A Matrix Calculator 188
10.6 Application: Moore-Penrose Inverse 190
10.7 Application: Curve Fitting 193
10.8 Elementary Matrices 198

11 Vectors 201
11.1 Linear Combinations 202
11.2 Linear Independence 203
11.3 The Range of a Matrix 207
11.4 The Kernel of a Matrix 208

12 Determinants 211
12.1 Permutations 211
12.2 Leibniz Formula for a Determinant 213
12.3 Laplace Expansion of a Determinant 214
12.4 Properties of Determinants 216
12.5 Determinants Using Row Echelon 216
12.6 Cramer's Rule 217
12.7 Application: Common Root of Polynomials 220
12.8 Application: Plane Through Three Points 222
12.9 Application: Sphere Through Four Points 224
12.10 Eigenvalues and Eigenvectors 226
12.11 Adjugate Matrix 227

13 Multivariable Algebra with Parameters 231
13.1 The Module 232
13.2 Application: Sums of Integer Powers 235
13.3 Application: Partial Fractions 238

Index 247

Python Essentials 1

Python is a powerful high level computer language with numerous features. In this chapter we consider those general aspects of the language that will be needed to achieve the mathematical goals of the book. More specialized features will appear later the text. While not necessary, the interested reader may wish to supplement the material in this chapter with one of the many excellent texts on Python.

1.1 Getting Started

There are two main ways to use Python. The first is to write a statement into a Python interpreter and press Enter; the result is then immediately displayed. This is useful as a quick learning tool or for testing parts of a long program. The second way is to use a text editor combined with a Python interpreter to create a text file (called a *script*) and then execute the file with a Run button. The file may be saved under a name with the .py extension. IDLE is the integrated editor which comes by default with Python. It include several features to make programming simpler, for example, automatic indenting and color coding. We shall use the script method throughout the book.

1.2 Functions and Methods

A *function* in Python is a named collection of statements executed together as a unit to perform a specific task. A function may be passed data (called *arguments* or *parameters*) and may return a value, but neither is necessary. A function runs only when it is *called*, that is, activated by its name, in a program. Functions are reusable so they can simplify code. Furthermore, splitting a program into a sequence of functions makes the overall logic clearer and debugging easier. The word *function*, as it is used here, should not be confused with its use in mathematics, although there are similarities: a mathematical function is a rule that takes an input and produces a unique output while a Python function is simply a named sequence of statements that carries out a task. It need not have input or output values.

There are two main types of functions in Python: built-in library functions and user-defined functions. An example of the former is the `print` function introduced in the next section. A user-defined function in Python is created using the keyword `def` followed by the name of the function and executable statements. Here is the general format:

```
def function_name(arguments):            # arguments are optional
    statements
    return value                          # return statement is optional
```

Note the colon and indentation. Python relies on the former to herald a forthcoming group of statements and the latter to indicate the grouping. The text following the hash sign `#` is intended to help explain the code; the Python interpreter skips over this. The use of liberally sprinkled comments can greatly enhance clarity and help reduce the chance of logical errors.

A *method* in Python is similar to a function in that is called to perform a particular task. Some methods may be passed parameters and may return values. Methods are called by placing a period between the object that the method is associated with and the method's name. We shall see examples of methods later. User defined methods are possible, but we won't need this feature.

1.3   Values and Variables

A *value* in Python is a data object such as an integer, decimal, or text. Values are characterized by their *types*. For example, the numbers 5, 0, and -2 belong the type *int* (integer). Numbers with decimal points belong the type *float* (floating point). Character sequences enclosed in quotation marks (double or single), as in "I am a string" or '3.14159', belong to the type *str* (string). Statements that are either true or false, have type *bool* (boolean). We discuss all of these types in more detail in the next few sections. The type *complex* refers to complex numbers, that is, expressions of the form $a + bj$, where a and b are real numbers and $j^2 = -1$. Python uses the letter j rather than the standard mathematical notation i to avoid conflict in electrical engineering contexts, where i stands for current. We shall not need this data type in our calculations as we consider complex numbers only within strings, freeing us to use the standard letter i for $\sqrt{-1}$.

1.3 Values and Variables

A *variable* is a name that references a value contained in computer memory. Python does not require that a variable be explicitly declared; it is created and its type decided the instant a value is assigned. For example, when the code var = 2.3 is executed, the value 2.3 is stored in computer memory and inherits the type float. The variable may then be used in code without destroying the data it is referencing. A variable name must obey certain rules. The name can only contain letters A–Z, a–z, digits 0–9, or underscores, and cannot start with a digit. Additionally, names are case sensitive; for example, V and v denote different variables. Finally, a name cannot be a Python *keyword*. These are reserved words that Python uses for built-in functions and variables, and for implementing certain tasks, such as branching or looping. The code help(''keywords'') disgorges Python's keywords.

The data type of a value or variable can be displayed by the built-in Python function type. Here are some examples using the print function mentioned earlier. We have also employed the Python convention of separating statements with semicolons. This is a handy space saver but should not be used if clarity is compromised.

```
Input:
# assign values to variables:
v1 = 37; v2 = 3.7; v3 = 'thirtyseven'; v4 = 3 < 7

# print data type:
print(type(37), type(3.7), type('thirtyseven'), type(3 < 7))

# print variable type:
print(type(v1), type(v2), type(v3), type(v4))

Output:
# data type of values:
<class 'int'> <class 'float'> <class 'str'> <class 'bool'>

# data type of variables:
<class 'int'> <class 'float'> <class 'str'> <class 'bool'>
```

The Python functions int(), float(), and str() may be used to convert values from one type to another. For example,

```
Input:
print(int(-3.7), float(37), float(3.7e-4), float(3.7e4), str(3.70))

Output:
-3  37.0  0.00037  37000.0  3.7
```

Some observations: The function int applied to a float chops off the fractional part of the number. For example, the statement print(int(3.9999),int(-3.9999)) prints the pair 3,-3. The notation 3.7e-4 is shorthand for the number 3.7×10^{-4} and calls for the decimal point to be shifted four places to the left. Similarly, 3.7e4 represents the number 3.7×10^4 and causes the decimal point to be moved four places to the right.

1.4 Numerical Operations

Python has several operations for evaluating arithmetic expressions. The most common of these are

+ addition
− substraction
* multiplication
/ division
** exponentiation

The operations follow the usual precedence rules in arithmetic: exponentiation first, then multiplication and division (no order of precedence between the two), and finally addition and subtraction (no order of precedence between the two). Parenthetical expressions are evaluated as they are encountered. Here are some examples:

```
Input:
print(5-6*3/2, 5-6/3*2, 5-6/2**3, 5-6**2/3, 6+7/8-(6+7)/8)

Output:
-4.0  1.0  4.25  -7.0  5.25
```

Python handles operations with both values and variables:

```
Input:
v1 = 2.3; v2 = 3.2           # assign values to variables v1 and v2
v = v1*v2 + v1/v2 - v1**v2   # assign an expression to v
print(v)

Output:
-6.293642707920499
```

1.4 Numerical Operations

There are three additional arithmetic operations of importance:

$$\begin{array}{ll} \texttt{abs()} & \text{absolute value function} \\ \texttt{//} & \text{floor division operator} \\ \texttt{\%} & \text{modulo operator} \end{array}$$

The first operator returns the absolute value of a number. For example, `abs(-2)` = 2. The floor division operator divides one real number by another and then rounds to the next lowest integer value. For example,

```
Input:
print(9//2, -9//2, 9//-2, -(9//2))

Output:
4 -5 -5 -4
```

The modulo operator returns the remainder r when a positive integer a is divided by a positive integer b, yielding the so-called *division algorithm*

$$a = q*b + r, \quad 0 \le r < b, \quad r = a \,\%\, b, \quad q = a//b$$

Here, a is called the *dividend*, b the *divisor*, and q the *quotient*. For example,

$$9 = (9//4)*4 + 9\,\%\,4 = 2*4 + 1.$$

Note that the remainder 1 may be expressed as $4*(9/4 - 9//4) = 4*(2.25 - 2)$. In general for positive integers a, b one has

$$a \,\%\, b = b*(a/b - a//b) \qquad (1.1)$$

Python takes this as the definition of the modulo operator for *all* integers a, b, $b \neq 0$. Try this with the function

```
def modulo_test(a,b):                        # enter integers a, b
    return a % b, b*(a/b - a//b)   # returns a pair of equal numbers
```

1.5 Boolean Operations

The Boolean data type `bool` refers to *conditional* statements, that is, statements that are either true or false. Python give these statements the keyword values `True` or `False`. The Python *comparison operators* are frequently used in this context:

$$
\begin{array}{ll}
== \text{ equal} & != \text{ not equal} \\
< \text{ less than} & <= \text{ less than or equal} \\
> \text{ greater than} & >= \text{ greater than or equal}
\end{array}
$$

One may use variables and operations in conditional statements:

```
Input:
v1 = (4 == 2*2)            # parentheses optional but add clarity
v2 = 6 > 3*2               # optional parentheses omitted here
v3 = 6 != 3*2                                          # and here
print(v1,v2,v3)
```
```
Output:
True False False
```

Since the comparison operator takes precedence over the assignment operator, the parentheses in the first statement of the input are not needed. For added clarity, however, it is recommended that they be used in contexts like this.

Conditional statements may be combined using the *logical operators* `and, or, not`. The operators are defined by the following rules, where A and B are conditional statements:

$$
\begin{array}{ll}
\text{A and B} & \text{true if and only if both A and B are true} \\
\text{A or B} & \text{true if and only if either A or B is true} \\
\text{not A} & \text{true if and only if A is false.}
\end{array}
$$

The above operations extend to more than two statements and follow the standard precedence rules: not first, then and, then or. Comparison operators are evaluated before logical operators. For example,

```
Input:
print(1 < 2 and not 3 < 4 or 5 < 6)
print(1 < 2 and not (3 < 4 or 5 < 6))
```
```
Output:
True
False
```

1.6 String Operations

The first statement is evaluated as

```
(True and not True) or True = (True and False) or True
                            = False or True
                            = True
```

and the second as

```
True and not(True or True)  = True and False
                            = False
```

While the first statement in the above input code is legitimate, for clarity it is better to write it as (1 < 2 and not 3 < 4) or 5 < 6.

1.6 String Operations

In this section we describe the more common string functions available in Python. Others appear in later applications.

Strings are joined together using the *concatenation operator* + (not to be confused with the addition operator).

```
Input:
print('fiddle' + 'faddle')
print('bibbity ' + 'bobbity ' + 'boo')          # spaces added

Output:
fiddlefaddle                                    # no spaces
bibbity bobbity boo   # spaces after bibbity and bobbity but not boo
```

The examples show that Python does not put spaces in strings unless it is directed to do so. It also prints strings without the quotes.

The *membership operator* in returns True if a specified substring is contained in a given string and False otherwise.

```
Input:
print('faddle' in 'fiddlefaddle', 'fuddle' in 'fiddlefaddle')

Output:
True False
```

A string is considered as a sequence of characters and as such may be referenced by their position or *index*. By convention, the first character of a string is located at index 0, the second at index 1, etc. We also say that an index *points to a character*. For example, the index 3 points to the character 'd' in the string 'abcde'. Indices may be negative, allowing a character to be referenced from the right. For example, the third character from the right in a string has index –3.

```
Input:
v = 'framosham'
print('framosham'[0],'framosham'[-3], v[2])

Output:
f h a
```

A *slice* or *substring* of a string may be accessed with the colon operator in the form [m:n]:

```
Input:
print('framosham'[3:8],'framosham'[5:9],'framosham'[5:])

Output:
mosha sham sham
```

Notice that in the notation [3:8] the letter at index 3 is included but the letter at index 8 is not. If you want a slice that includes the last character of the string you can use a "fictitious" index which is one more than the index of the last character, or you can use colon notation such as [j:]. For example, 'drizzle'[2:7]='drizzle'[2:]='izzle' You can also use [:j] as a substitute for [0:j]. Each of these yields the *j* characters from the beginning of a string up to, but not including, the jth character. For example, 'drizzle'[0:5]='drizzle'[:5]='drizz'.

Here's a script defining a function that uses the slice method to switch the first and last letters of a given string. The first statement uses the length function len, which returns the length of a string.

```
def switch_first_and_last(string):
    L = len(string)                    # length of string
    first = string[0]                  # first character in string
    last  = string[L-1]                # last character in string
    middle = string[1:L-1]     # characters at positions 1,2,... L-2
    return last + middle + first       # switch first and last
```

1.6 String Operations

```
Input:
print(switch_first_and_last('read'))
```
```
Output:
dear
```

The Python `replace` method takes every occurrence of a substring in a given string and replaces it by another string. In the following example, every occurrence of the substring ''doo'' is replaced by the string ''noo''.

```
Input:
farmsound = "cockadoodledoo"                        # given string
newfarmsound = farmsound.replace("doo","noo")       # new string
print(farmsound + ',', newfarmsound)                # comma separation
```
```
Output:
cockadoodledoo, cockanoodlenoo
```

Notice that the original string ''cockadoodledoo'' is unchanged. This is the case for all string methods: they just return new values. Note also that we have concatenated a comma to the string variable `farmsound` in the print function to indicate separation of the items.

The Python method `lower` converts all letters of a string to lower case. Its analog `upper` does the reverse. The methods leave non-letters untouched.

```
Input:
print('frazzle'.upper())                 # invoke upper method
print('2BEES OR NOT 2BEES'.lower())      # invoke lower method
```
```
Output:
FRAZZLE
2bees or not 2bees
```

The `count` method returns how many times one string appears in another, either with or without index constraints:

```
Input:
st = "framosham"
n1 = st.count("am")          # total number of am's in str
n2 = st.count("am",7)        # number of am's starting at index 7
n3 = st.count("am",2,4)      # number of am's from 2 to 3 inclusive
n4 = st.count("am",8)        # number of am's starting at index 8
print(n1, n2, n3, n4)
```

```
Output:
2 1 1 0
```

The find method returns the index in a string of the first occurrence of a given substring, returning −1 if the substring is not found. Index constraints may be imposed as in the method count and follow the same rules.

```
s = 'framosham'
s1 = st.find('am')                  # index of first 'am' in str
s2 = st.find('am',7)        # index of first 'am' starting at index 7
s3 = st.find('am',2,4)  # index of first 'am' between 2, 3 inclusive
s4 = st.find('am',8)        # index of first 'am' starting at index 8
print(s1, s2, s3, s4)

Output:
2 7 2 -1
```

A string is *immutable*, that is, the characters cannot be changed. For example, the following code does not produce the desired string `faddle` but rather throws an error. We show in Sect. 1.8 how to get around this feature.

```
Input:
'fiddle[1]' = 'a'

Output:
TypeError: 'str' object does not support item assignment
```

1.7 ASCII Functions

ASCII is an abbreviation for "American Standard Code for Information Interchange." The code represents English letters and other characters by various symbols. We shall only need the integer code for upper case letters, lower case letters, and digits:

- A - Z: 65 - 90
- a - z: 97 - 122
- 0 - 9: 48 - 57.

The Python function ord returns the ASCII code for a given character. The Python function chr does the reverse: it takes an ASCII code number and returns the corresponding character.

```
Input:
print(ord("a"), ord("e"), ord("i"), ord("o"), ord("u"))
print(chr(97), chr(101), chr(105), chr(111), chr(117))
```
```
Output:
97 101 105 111 117
a e i o u
```

1.8 Lists

A list in Python is an indexed sequence of items separated by commas and enclosed by square brackets. It is one of the most-used data structures in Python, and we shall have many occasions to use it. Part of its usefulness derives from its great versatility, including the property that items may be of different types. For example, the following code defines a list variable A containing a string, a float, an integer, a boolean expression, and a type command.

```
Input:
A = ['frobish', -4.2, 11, 4 == 5, type(3+2j)]
print(A)
```
```
Output:
['frobish', -4.2, 11, False, <class 'complex'>]
```

A list can contain other lists:

```
Input:
B = [A, 'I am not a list']
print(B)
```
```
Output:
[['frobish',-4.2,11,False,<class 'complex'>],'I am not a list']
```

Like strings, lists may be concatenated using a plus sign:

```
Input:
print(["alpha","beta"] + [1,2,3])

Output:
['alpha, 'beta',1, 2, 3]
```

As with strings, the members of a list may be referenced using indices which start at 0. For example,

```
Input:
C = [['a','b','c'], 3,2,1]
print(C[0], C[1], C[0][2])

Output:
['a', 'b', 'c'] 3 c
```

The `index` method returns the index of the first occurrence of an item in the list:

```
Input:
print(['a','b','c','d'].index('c'))

Output:
2
```

The `slice` function for lists works the same way as for strings, obeying the same rules.

```
Input:
A = ['frobish', -4.2, 11, 4 == 5, type(3+2j)]
print(A[1:3])              # print items with indices 1 and 2
print(A[2:])               # print items from index 2 on
print(A[:3])               # print items with indices 0,1 and 2

Output:
[-4.2, 11]
[11, False, <class 'complex'>]
['frobish', -4.2, 11]
```

1.8 Lists

The `join` method merges into a string the members of a list, with or without a delimiter.

```
Input:
string = ['aaa', 'bbb', 'ccc']
list1 = ''.join(string)         # join items in list without delimiter
list2 = ','.join(string)        # join items with comma delimiter
print(list1,' ',list2)          # insert a space for clarity

Output:
aaabbbccc   aaa,bbb,ccc
```

The `split` method takes a string and returns a list of its members separated at each occurrence of a given string character. The separators are eliminated in the splitting process.

```
Input:
river = 'Mississippi'
print(river.split('i'))         # separate at the i's
print(river.split('s'))         # separate at the s's

Output:
['M', 'ss', 'ss', 'pp', '']              # no i's
['Mi', '', 'i', '', 'ippi']              # no s's
```

The `list` function converts a string into a list:

```
Input:
river = 'Mississippi'
river_list = list('Mississippi')
print(river_list)

Output:
['M','i','s','s','i','s','s','i','p','p','i']
```

As with strings, the operator `in` returns True if a given item is contained in a list:

```
Input:
A = ['frobish', -4.2, 11, 4 == 5, type(3+2j)]
print('frobish' in A, 'frubish' in A)

Output:
True False
```

The append method adds another item to a list. The `insert` method places a new item at a designated index.

```
Input:
E = []                                       # start with an empty list
E.append('Afghanistan')                                    # first member
print(E)
E.append('Stand')                                         # second member
print(E)
E.insert(1,'Banana')         # insert 'Banana after 'Afghanistan'
print(E)
Output:
['Afghanistan']
['Afghanistan', 'Stand']
['Afghanistan', 'Banana', 'Stand']
```

Items in a list may be changed simply by overwriting. Thus, in contrast to strings, lists are *mutable*.

```
Input:
E = ['Afghanistan', 'Banana', 'Stand']
E[1]= 'Bandana'                                     # overwrite 'Banana'
print(E)
Output:
['Afghanistan', 'Bandana', 'Stand']
```

Items are eliminated by the `remove` method. The `clear` method removes *all* items.

```
Input:
E = ['Afghanistan', 'Banana', 'Stand']
E.remove('Afghanistan')
print(E)
E.clear()                                         # remove everything
print(E)
Output:
['Banana', 'Stand']
[]                                                        # empty list
```

1.9 Tuples

A tuple is an ordered collection of items separated by commas and enclosed by parentheses. It is initialized in the same way as a list, and members of a tuple can be accessed by using the square bracket index method, exactly as in lists. However, unlike a list, a tuple is immutable, that is, it cannot be modified. To get around this, one can convert a tuple to a list, change the list, and then convert the resulting list to a tuple:

```
Input:
t = (1,2,3,4,5)               # create a tuple variable
print(t)
s = list(t)                   # convert it to a list
s[3] = 7                      # change the 4 to a 7
t = tuple(s)                  # convert list to tuple
print(t)

Output:
(1, 2, 3, 4, 5)
(1, 2, 3, 7, 5)
```

Tuples are used much less frequently than lists. They are useful in situations where it is crucial that a particular sequence of items be preserved. We shall have little if any occasion to use them.

1.10 Sets

The notion of set in Python corresponds closely to its definition in mathematics: a collection of items separated by commas and enclosed by braces. In contrast to lists and tuples, however, sets are not indexed, do not have duplicate members and are not ordered. But one can determine membership and append or remove members of a set directly without using indices.

```
Input:
s = {1, 2, 3, 'x', 'y'}       # initialize set s
print(2 in s)                 # check membership
print(5 in s)
s.add('b')                    # add another member
print(s)
s.remove('y')                 # get rid of 'y'
print(s)

Output:
True
```

```
False
{1, 2, 3, 'b', 'x', 'y'}
{1, 2, 3, 'b', 'x'}
```

The Python function `set` may be used to convert lists and strings into sets. It may also be used in conjunction with the `list` function to remove duplicates from a list.

```
Input:
print(list(set('Mississippi')))

Output:
['p', 'M', 's', 'i']
```

1.11 Dictionaries

Dictionaries are similar to lists, the essential difference being that items are appended in the form `key:value`, where the key is used to find the value as in lists (slicing is not available). Entries can be deleted using the `del` statement and revised or added items using the assignment operator. Here is an example that gives the test scores of several students.

```
Input:
sample_dictionary = {'Betty':82,'Ralph':67,'Aaron':85,'Sally':91}
print(sample_dictionary['Sally'])          # reveal Sally's score
del sample_dictionary['Ralph']             # Ralph dropped course
sample_dictionary['Betty'] = 84            # give Betty 2 more points
sample_dictionary['Bruce'] = 72            # add Bruce's late test score
print(sample_dictionary)

Output:
91                                         # Sally's score
{'Betty': 84, 'Aaron': 85, 'Sally': 91, 'Bruce': 72}
```

1.12 The If Elif Else Statement

The `if elif else` statement is an example of a *conditional statement*. Such statements enable the path of a program to branch to other statements depending on conditions that occur during run time. These conditions are usually in the form of Boolean expressions and

1.12 The If Elif Else Statement

typically involve comparison and logical operators, discussed earlier. The statement has the following general form:

```
if condition1:
    statements1 # statements indented by the recommended 4 spaces
elif condition2:
    statements2
    ...
elif conditionk:
    statementsk
else:
    statements
```

There may be arbitrarily many `elif` statements or none at all. The code executes as follows: if `condition1` is true, then `statements1` are executed, the remaining `elif` and `else` statements, if any, are ignored, and control passes to the statement following the code block. On the other hand, if `condition1` is false, then `condition2` is tested. The process continues in this manner through the kth statement. The `else` statement, which is optional, is executed if conditions 1 to k are false.

Here's an example of a function that determines which of three distinct numbers is between the other two.

```
def in_between(a,b,c):
    if (c < a and a < b) or (c > a and a > b):
        return a
    elif (c < b and b < a) or (c > b and b > a):
        return b
    elif (a < c and c < b) or (a > c and c > b):
        return c
    else:
        return "There is no such number."
```

--------------------------- Sample Run ---------------------------
Input:
print(in_between(3,1,2), in_between(3,2,2))

Output:
2 There is no such number.

1.13 The While Loop

A loop in a program causes a sequence of instructions to execute repeatedly until a given condition changes, whereupon the loop exits. Such statements are typically used with comparison and logical operators. Loops are indispensable in calculations that require a large number of repeated calculations. There two main types of loops in Python: the `while` loop and the `for` loop. We consider the former in this section, the latter in the next.

The `while` loop continues until a condition becomes `False`. Here is the form that we shall need.

```
while condition:            # continue while condition is true
    statements              # statements are indented
```

The following example takes a string and reverses adjacent pairs of characters of the string, leaving the last one untouched if the number of characters is odd.

```
def reversepairs(str):                      # input a string str
    L = len(str)
    idx = 0                                 # index for string
    revpairs = ''                           # create an empty string
    while idx <= L-2:   continue
        revpairs = revpairs + str[idx+1]    # reverse characters
        revpairs = revpairs + str[idx]
        idx = idx+2                         # increment index by 2
    if L%2 != 0:                            # if L not even
        revpairs = revpairs + str[L-1]      # odd number of letters
    return revpairs

--------------------------- Sample Run ---------------------------
Input:
print(reversepairs('123456')   # last iteration at idx = 4,digit = 5
print(reversepairs('1234567')) # last iteration at idx = 5,digit = 6

Output:
214365 2143657
```

Here's example of a function that calculates the smallest integer n for a sum of the form

$$1 + \frac{1}{2} + \frac{1}{3} + \cdots + \frac{1}{n}$$

to exceed a given number U (upper bound). The sums grow larger and larger but do so very slowly. For example, running the function reveals that 100,210,581 terms are needed for the sum to exceed 19, and an additional 172,190,019 terms are required to get past 20.

1.13 The While Loop

```
def num_reciprocals(U):
  s = 0; n = 0                    # initialize sum and counter
  while s <= U:                   # stop loop when sum > U
    n = n + 1                     # increment n
    s = s + 1/n                   # update sum with new term
  return n
-------------------------- Sample Run ---------------------------
Input:
print(num_reciprocals(10))

Output:
12367
```

You can skip some iterations in a while loop by using the continue statement. For example:

```
def num_odd_reciprocals(U):
  s = 0; n = 0                    # initialize sum and counter
  while s <= U:                   # stop loop when sum > U
    n = n + 1                     # increment n
    if n%2 == 0: continue         # skip even numbers
    s= s + 1/n                    # update sum with new term
  return n
-------------------------- Sample Run ---------------------------
Input:
print('n = ', sum_of_odd_reciprocals(10))

Output:
n =  136200301
```

You can get out of a while loop by using the break statement. Here's an example that takes a list of numbers and returns the product of its reciprocals. The loop breaks if it encounters a zero, in which case the function returns None.

```
def invert_and_multiply(numberstring):
    numberlist = numberstring.split(',')   # make a list from string
    L = len(numberlist)
    prod = 1; i = 0                         # initialize product and index

    while i < L:
        if numberlist[i] == '0':
            break
        reciprocal = 1/float(numberlist[i])
```

```
        prod = prod*reciprocal
        i += 1                          # simplified notation for i = i+1
    if i == L:                          # made it through ok?
        return prod                                                 # yep

--------------------------- Sample Run ---------------------------
Input:
numberstring = '2,-1,3.3,4.7,5.2,.00003'
print(invert_and_multiply(numberstring))
------------------------------------------------------------------
Output:
-206.64914281935557
------------------------------------------------------------------
```

One can replace the `break` statement by `return`, obviating the last two lines of code.

1.14 The For Loop

The `for` loop iterates through a collection objects. Again, indentation is required.

```
------------------------------------------------------------------
for item in collection:                  # a list or string
    statements   # execute until no more items left in collection
------------------------------------------------------------------
```

Here's an example that takes a string and returns a string of its vowels and a string of its consonants. Note that the function returns both strings at once, feature that we avail ourselves of throughout the text.

```
------------------------------------------------------------------
def separate_vowels_consonants(instring):
    vowels = ''                         # initialize with null string
    consonants = ''
    for ch in instring:
        if ch in 'aeiou':
            vowels = vowels + ch  # concatenate to the vowels string
        if ord(ch) in range(97,123) and ch not in 'aeiou':
            consonants = consonants + ch    # ditto for consonants
    return vowels, consonants     # return both strings in a pair

--------------------------- Sample Run ---------------------------
Input:
instring = 'honorificabilitudinatatibus'
print(separate_vowels_consonants(instring))
------------------------------------------------------------------
Output:
('ooiiaiiuiaaiu', 'hnrfcbltdnttbs')     # strings returned in a pair
------------------------------------------------------------------
```

1.14 The For Loop

One can also use the `range` statement in place of collections, specifying start and end constraints:

```
Input:
for k in range(2,5):                                    # 2 <= k < 5
    print(k, end=' ')

Output:
2 3 4
```

Note that in the range statement the lower number is included but the upper number is not. Note also use of the `end` feature in the print function, which enables one to print in a single line with an intervening symbol, in this case a space.

One can omit iterations with a skip value that replaces the default skip of 1. The following example prints every third value of k, starting with $k = 2$ and ending $k = 11$. The skip value is the last digit in the `range` statement, in this case 3.

```
Input:
for k in range(2,12,3):
    print(k, end=" ")        # print every third k starting with 2

Output:
2 5 8 11
```

The skip value can be negative, causing the iteration to reverse.

```
Input:
# print every third k starting from the end 12. Stop before 2
for k in range(12,2,-3):
    print(k, end=" ")

Output:
12 9 6 3
```

The `continue` and `break` are available and are used exactly as in `while` loops:

```
Input:
for k in "Mississippi":
    if k == "s" or k == "M": continue       # skip these letters
    if k == "p": break
    print(k,end=" ")
```

```
Output:
i i i
```

You can loop over a dictionary as well:

```
Input:
gradebook = {"Betty": 82, "Ralph": 67, "Aaron": 85, "Sally": 91}
for student in gradebook:
    print(student, end=" ")          # print names on a single line
print('\n')                          # skip a line
for grade in gradebook:
    print(gradebook[grade],end=" ") # print grades on a single line
Output:
Betty Ralph Aaron Sally
                                                    # line skipped
82 67 85 91
```

Here we have used the *newline character* '\n' to skip a line.

1.15 Recursion

The word "recursion" in programming refers to a procedure that calls itself. It is similar to looping in that statements are executed repeatedly. Iterations in a loop cease when a certain condition is satisfied; recursion ends when a *base case* or *terminating condition* is met.

The factorial function provides a simple example of recursion. The diagram in Fig. 1.1 indicates the various recursion levels undergone by the process. Note that multiplication is delayed (the "ladder" is descended without the operations being performed) until the

Fig. 1.1 Factorial recursion

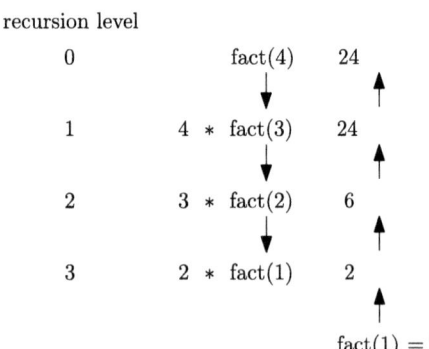

1.15 Recursion

base case $n = 1$ is reached, whereupon the ladder is ascended, the delayed multiplication performed at each "rung" of the climb.

```
def factorial(n)
    f = 1                                        # value for n = 1
    if n > 1:
        f = n * factorial(n - 1)                 # do the recursion
    return f
```

```
---------------------------- Sample Run ----------------------------
Input:
print(factorial(50))
```
Output:
30414093201713378043612608166064768844377641568960512000000000000

Here's another recursive function. It takes a starting value $a_1 = 1$ and a positive number r and generates a sequence $\{a_n\}$ using the formula

$$a_n = \frac{1}{2}\left(a_{n-1} + \frac{r}{a_{n-1}}\right)$$

The sequence gets closer and closer to $\sqrt{r}$.

```
def square_root(r,n):
    a = 1                                        # return 1 if n = 1
    if n > 1:                # continue calling function until n = 1
        a = square_root(r,n-1)                   # do the recursion
        a = (a + r/a)/2
    return a
```

```
---------------------------- Sample Run ----------------------------
Input:
r = 2                                  # generate square root of 2
n = 100  # no. of iterations    (more gives a better approximation)
s = square_root(r,n)                   # nth approximation to square root
print(s-2**(1/2))                      # check against Python's square root
```
Output:
-2.220446049250313e-16   # very small difference

The reader will find simple ways to achieve the same output in these examples using a loop. However, as examples throughout the book demonstrate, recursion may be the only practical solution to an iterative task.

1.16 Modules

A Python module is a text file that contains functions, variables, runnable code, and other items. A good example is the module `math`, which contains trig functions, power function, logarithmic functions, mathematical constants such as π and Euler's constant e. We shall occasionally use this module. Modules can also be created. Indeed, the code in this chapter has been assembled in the module `Essentials.py`. We shall create additional modules throughout the text.

General Tools 2

In this chapter we develop functions that will be used in a variety of contexts throughout the book. They deal mainly with arithmetic, algebraic, and logical expressions. The reader may wish to skim the chapter first, coming back to a particular section as needed. The functions in this chapter comprise the module `Tools.py`. The module is headed by the following statements.

```
------------------------- Tools.py --------------------------
global upper, lower, letters, numeric
upper = 'ABCDEFGHIJKLMNOPQRSTUVWXYZ'
lower = 'abcdefghjklmnopqrstuvwxyz'       # note the missing 'i'
letters = upper+lower
numeric = '.0123456789i'
-------------------------------------------------------------
```

The term `global` refers to variables that are available to all functions in the module.

2.1 Mathematical Expressions

Many of the functions in the chapter deal with mathematical expressions. These fall into three main categories: *arithmetic expressions*, *algebraic expressions*, and *logical expressions*. An arithmetic expression is a combination of sums, differences, products, quotients, and powers of *numerics*: integers, decimals, fractions and the imaginary number $i = \sqrt{-1}$. We call[1] Here is an example of an arithmetic expression:

```
1.2*4.3 + 2.1*(3.1/4.2 - 5.3i)^11/(6.4i + (7+i)^20)^(-3).
```

Note that the expression is written using standard typewriter mathematical notation. In particular, an asterisk denotes multiplication and the caret symbol denotes exponentiation. We use the caret symbol instead of Python's double asterisk for compactness, better readability, and deference to mathematical convention. Single asterisks between terms may be omitted if no ambiguity results. For example, the second asterisk in the above expression may be removed but not the first, although one could also write (1.2)(4.3) for that term. Parentheses in exponents, negative or positive, are optional unless there is ambiguity. Parentheses should be used when in doubt.

An algebraic expression is similar to an arithmetic expression except that it may contain letters representing variables or parameters, as in

```
(3x^2 + 7.1A + 5i)/(2y + 1)^123 - 2.3z.
```

The rules governing arithmetic expressions apply here as well. Variables may have subscripts, as in x22. The only requirement on subscripts is that pairs such as x and its subscripted relative x22 may not appear in the same expression. The letter i is deemed a numeric and so is considered neither a variable nor a parameter.

A logical expression contains letters and the operations

```
p+q, pq = p*q, p+q, p->q, p<->q, p'
```

These will be explained in detail in Chap. 3. An example of such an expression is (q + r')<->(rs -> t)

All mathematical expressions in the book are entered as strings and manipulated symbolically.

2.2 Category Functions

The following functions determine if a character is of a particular type, as described in the above module heading. The first returns False if a letter is found in an expression.

---

[1] Recall that we use the standard mathematical symbol i rather than Python's symbol j.

```
def isarithmetic(expr):
    for ch in expr:
        if ch in letters:
            return False
    return True

def isnumeric(ch):
    return ch in numeric

def isletter(ch):
    return ch in letters

def isupper(ch):
    return ch in upper

def islower(ch):
    return ch in lower
```

2.3 Index Functions

The function move_past(expr,start,string) takes an expression, a starting index, and a string of characters, and moves the index from left to right through the expression until it no longer points to a member of the string. It then returns that index. For example, if expr ='abc1984xyz', start = 4, (pointing to the digit 9), and 'string = 1234567890', then the function returns 7, the index of the character 'x'.

```
def movepast(expr,start,string):
    idx = start                       # index of character in expr
    while idx  < len(expr):
        if expr[idx] not in string:
            break
        idx+=1
    return idx    # idx now points to the right of last string symbol
```

The function move2rparen(expr,start) takes an expression and an index pointing to a left parenthesis in the expression and returns the index of the matching right parenthesis by counting instances of both. For example, if expr ='(1+(2+(3+4)))' and start = 3 (the second left parenthesis) then the function returns 11 (the second right parenthesis). At this index the numbers of left and right parentheses are equal.

```
def move2rparen(expr,start):
    # returns index of matching right paren; start is at '('
    numleft = 0; numright = 0
    idx = start
    while idx < len(expr):
        ch = expr[idx]
        if ch == '(':
            numleft += 1      # shorthand for numleft = numleft + 1
        if ch == ')':
            numright += 1
        if numleft == numright:
            break
        idx += 1
    return idx                            # idx now at matching ')'
```

2.4 Retrieving Symbols

The functions in this section return various types of symbols in an expression. The first considers only lower case letters unequal to i, these customarily used for variables in an expression. The second function considers lower *and* upper case letters, the latter customarily used for parameters. It uses the Python function `sorted`, which sorts the variables alphabetically for easy reference.

```
def get_var(expr):         # returns the first lower case letter != i
    for ch in expr:
        if ch in lower: return ch
    return ''       # no lower case letters found; return null string

def get_vars(expr):
    varlist = []
    j=0
    while j < len(expr):
        if expr[j] in letters:
            start = j
            end = movepast(expr,start+1,'1234567890')  # subscripts?
            var = expr[start:end]                       the variable
            varlist = varlist + [var]              # attach to list
            j = end
        else:
            j += 1
    varlist = sorted(list(set(varlist)))   # kill duplicates and sort
    varstring = ''.join(varlist)                     # string form
    return varlist,varstring                 # return list and string
```

```
def get_lower(expr):
    # gets all lower case letters (except i)
    variables = []
    for ch in expr:
        if ch in lower:
            variables = variables + [ch]
    if variables == []:
        return ''
    return ''.join(sorted(list(set(variables))))     # return string

def get_upper(expr):
    # gets all upper case letters
    variables = []
    for ch in expr:
        if ch in upper:
            variables = variables + [ch]
    if variables == []:
        return ''
    return ''.join(sorted(list(set(variables))))
```

```
-------------------------- Sample Run ---------------------------
Input:
expr = '7x11 +8x12+ 9x13+10x14+11ABcd'
print(get_vars(expr)[0])
print(get_lower(expr))
print(get_upper(expr))
-----------------------------------------------------------------
Output:
['A', 'B', 'c', 'd', 'x11']
cdx
AB
-----------------------------------------------------------------
```

2.5 Preparing Expressions

Several functions are needed to put a given expression into a form suitable for computations. The first such function, insert_asterisks, takes a mathematical expression and places asterisks between various symbols so that programs can deal more simply with multiplication operations. The symbols are numerics and upper and lower case letters (with or without subscripts).

```
-----------------------------------------------------------------
def insert_asterisks(expr,varbs):       # varbs = allowable variables
    expr = expr.replace(' ','')         # remove white (blank) space
    expr = expr.replace(')(',')*(')     # insert between parens
    for v in varbs:                     # insert between paren and var
        expr = expr.replace(')'+v,')*'+v)
        expr = expr.replace(v+'(',v+'*(')
```

```
        for n in numeric:              # insert between paren and numeric
            expr = expr.replace(')'+n,')*'+n)
            expr = expr.replace(n+'(',n+'*(')
        for m in varbs:                 # insert between variables
            for n in varbs:
                expr = expr.replace(m+n,m+'*'+n)
        for v in varbs:         # insert between variable and constants
            for n in numeric:
                expr = expr.replace(n+v,n+'*'+v)
        for v in varbs: # insert between accent and variable (for logic)
            expr = expr.replace("'"+v,"'*"+v)
        expr = expr.replace("'(","'*(")                 # for logic
        return expr

------------------------- Sample Run -------------------------
Input:
expr = 'Ayz34v+uB(3.7i+x2F11G_12)^w5'
varbs = get_vars(expr)[0]
print(varbs)
print(insert_asterisks(expr,varbs))

expr = "(p->r'q)(q->(r+p'))"
varbs = get_vars(expr)[0]
print(varbs)
print(insert_asterisks(expr))
--------------------------------------------------------------
Output:
['A', 'B', 'F11', 'G12', 'u', 'v', 'w5', 'x2', 'y', 'z34']
A*y*z34*v+u*B*(3.7i+x2*F11*G12)^w5

['p', 'q', 'r']
(p->r'*q)*(q->(r+p'))
--------------------------------------------------------------
```

The function fix_operands ensures that plus and minus signs have two operands. Specifically, it takes a numeric with a lonely plus or minus sign and prefaces the sign with a zero so that in calculations all operations are binary. For example, (-5-7) is converted to (0-5-7).

```
--------------------------------------------------------------
def fix_operands(expr):
    if expr[0] == '-':              # if '-' at beginning, then
        expr = '0-' + expr[1:]                      # insert '0'
    if expr[0] == '+':                       # similarly for '+'
        expr = '0+' + expr[1:]
    expr = expr.replace('(-', '(0-',)
    expr = expr.replace('(+', '(0+',)
    return expr
--------------------------------------------------------------
```

2.5 Preparing Expressions

Certain functions may produce unwanted double signs during run time. The function `fix_signs(expr)` fixes this as well as related problems.

```
def fix_signs(expr):
    if expr == '' or expr == '0': return expr
    expr = expr.replace('(-)', '-')
    expr = expr.replace('(+)', '+')
    expr = expr.replace('++','+')
    expr = expr.replace('-+','-')
    expr = expr.replace('+-','-')
    expr = expr.replace('--','+')
    expr = expr.replace('+=','=')
    expr = expr.replace('-=','=')
    expr = expr.replace('=+','=')
    expr = expr.replace('=0-', '=-')
    expr = expr.replace('=0+', '=')
    # remove beginning and signs:
    if expr[0] == '+': expr = expr[1:]
    if L>1 and expr[L-1] in '+-': expr = expr[:L-2]
    return expr
```

The following function attaches '^1' to a variable in `expr` missing a exponent. This makes the task of extracting exponents uniform. The function ignores variables not specified in `var_list`. A letter and a subscripted version may not both appear in `var_list`

```
def attach_missing_exp(expr,var_list):
expr = expr.replace(' ','')
for v in var_list:
    expr = expr.replace(v,v + '^1')
 return  expr.replace('^1^','^')
```

```
-------------------------- Sample Run ---------------------------
Input:
expr = 'x123+y^456z789'
var_list = ['x123','y','z789']
print(attach_missing_exp(expr,var_list))
-----------------------------------------------------------------
Output:
x123^1+y^456z789^1
-----------------------------------------------------------------
```

2.6 Extracting Character Groups

The following functions extract various character sequences from an expression. The first of these takes an expression `expr`, an index `start` pointing to the beginning of a desired sequence in the expression, and the allowable characters in the sequence, and returns the sequence and the index that points one position beyond the sequence. The remaining functions are special cases of the first.

```
def extract_sequence(expr,start,characters):
    end = movepast(expr,start,characters)
    return expr[start:end], end         # end is one past sequence

def extract_numeric(expr,start):
    return extract_sequence(expr,start,'.0123456789i')

def extract_integer(expr,start):
    return extract_sequence(expr,start,'0123456789')
```

--------------------------- Sample Run ---------------------------
```
Input:
expr = '3-4+.5432ia5+6'

start = 4                               # index of decimal point
print(extract_numeric(expr, start))

start = 10                              # index of 'a'
print(extract_var(expr, start))
```

```
Output:
('.5432i', 10)
('a5', 12)
```

The next functions are similar in spirit. The function `extract_paren` uses `move2rparen` to extract a parenthetic expression. The function `extract_exp` takes an expression and index pointing to the symbol '^' in the expression and returns the exponent that follows as well as the index immediately after the exponent. Allowable characters are an initial minus sign, digits 0–9, and the letter *t*, which stands for transpose discussed in a later chapter. Parentheses around the exponent are optional but should be used if lack thereof makes for ambiguity.

```
def extract_paren(expr,start):          # index start is at '('
    end = move2rparen(expr,start)       # end at ')'
    end += 1                            # one past ')'
    return expr[start:end], end         # return '(...)', index
```

2.7 Inserting and Replacing Characters

```
def extract_exp(expr,start):            # index start is at '^'
    start += 1                           # move one past '^'
    k = start
    if expr[k] == '(':                  # parenthetic exponent?
        return extract_paren(expr,start) # return group in parens
    if expr[k] == '-':                  # negative exp?
        k+=1                             # skip past
    string = '1234567890t'   # allowable characters (t = transpose)
    end = movepast(expr,k,string)       # end is one past exponent
    return expr[start:end], end

--------------------------- Sample Run ---------------------------
Input:
expr = '3+(7-i+x)^(-54t)+11'
start = 9                                # index of '^'
print(extract_exp(expr,start))

expr = '3+(7-i+x)^-54t+11'
start = 9
print(extract_exp(expr,start))

expr = '(1+(2+((3+4)+5))+6)'
start = 7                                # index of 4th left paren
print(extract_paren(expr,start))
------------------------------------------------------------------
Output:
('(-54t)', 16)
('-54t', 14)
('(3+4)', 12)
------------------------------------------------------------------
```

2.7 Inserting and Replacing Characters

The functions in this section insert one string into another or replace part of the string with another string. The first such function places parentheses around expressions preceded by a negative sign or containing a fraction symbol '/'. For example, add_parens('-3') returns '(-3)' and add_parens('1/2') returns '(1/2)'.

```
------------------------------------------------------------------
def add_parens(expr):
    if expr == '' or expr == '-' :
        return expr
    if expr[0]  == '(':
        return expr       # already has parens
    if '/' in expr or '+' in expr[1:] or '-' in expr[1:]:
     return '(' + expr + ')'
    return expr
------------------------------------------------------------------
```

The function `replace_string(string,replacement,idx)` removes the portion of `string` of length `len(replacement)` that starts at index `idx` and replaces it with the string `replacement`. It returns the new string and the index that points immediately after the replacement.

```
def replace_string(string,replacement,idx):
    L = len(replacement)
    left = string[:idx]              # portion of string up to idx-1
    right = string[idx+L:]           # portion after idx+L-1
    return left + replacement + right, idx + L

------------------------- Sample Run -------------------------
Input:
print(replace_string('abcdefg','xy',0))      # replace ab with xy
print(replace_string('abcdefg',xy,3))        # replace de with xy

Output:
('xycdefg', 2)                               # index points to c
('abcxyfg', 5)                               # index points to f
```

The following function inserts a string into an expression at a specified index without removing any part of the original string. It also returns the index that points to the character immediately following the insertion.

```
def insert_string(expr,insertion,idx):
    outstring = expr[:idx] + insertion + expr[idx:]
    return outstring, idx + len(insertion)
------------------------- Sample Run -------------------------
Input:
print(insert_string('abcd','xy',0))      # inserts 'xy' before 'abcd'
print(insert_string('abcd','xy',2))      # inserts 'xy' after 'b'

Output:
('xyabcd', 2)
('abxycd', 4)
```

2.8 List Functions

The function `string2table` takes a string with certain delimiters and returns a list, which we shall call a *table*. The comma-separated portions of the string form sublists, which we shall call *rows*. The rows have lengths determined by semicolon delimiters. The function

2.8 List Functions

provides a convenient way for entering tables, and in particular matrices, into functions that require such lists.

```
def string2table(s):
    s = s.replace(' ','')                    # remove extra space
    if ';' not in s:
        row = s.split(',')                   # no semicolon
        table = [row]
        return table                         # single row in list
    rows = s.split(';')                      # split into lists at semicolon
    table = []
    for row in rows:                         # split row entries at commas
        rowlist = row.split(',')
        table.append(rowlist)
    return table

------------------------- Sample Run -------------------------
Input:
s = '1,2;3;4,5,6'                            # 3 rows
print(string2table(s))
s = '1;2;3;4;5;6'                            # 6 rows
print(string2table(s))
s = '1,2,3,4,5,6'                            # 1 row
print(string2table(s))
--------------------------------------------------------------
Output:
[['1', '2'], ['3'], ['4', '5', '6']]
[['1'], ['2'], ['3'], ['4'], ['5'], ['6']]
[['1', '2', '3', '4', '5', '6']]
--------------------------------------------------------------
```

The following function does the reverse of `string2table`, converting a table into a string.

```
def table2string(T):
    # takes a table and prints a comma\semicolon delimited string
    if isinstance(T[0],str):                 # single string
        return ','.join(T)
    s = ''
    for row in T:
        rowstr = ','.join(row)
        s = s + ';' + rowstr
    return s[1:]

------------------------- Sample Run -------------------------
Input:
T = [['1','2','3'],['4','5']]
print(table2string(T))
T = ['1','2','3','4','5','6']
```

```
print(table2string(T))
```
```
Output:
1,2,3;4,5
1,2,3,4,5,6
```

The function `flatten_double_list(dlist)` takes a double list (list of lists) and returns a single list.

```
def flatten_double_list(dlist):
    slist = []
    for d in dlist:
        for s in d:
            slist.append(s)
    return slist
```
---------------------------- Sample Run ----------------------------
```
Input:
dlist = [['1','2','3'],['4','5','6'],['7','8'],['9']]   # double list
print(flatten_double_list(dlist))
```
```
Output:
['1', '2', '3', '4', '5', '6', '7', '8', '9']
```

The function `zero_list(n)` returns a list of n zeros in string form.

```
def zero_list(n):
return ['0' for k in range(n)]
```

The function `copy_list(L)` takes a list L of any complexity and returns independent copy of the list. This means that any changes made to the copy will not affect the original list L. This is in contrast to what happens when one sets a variable K to L: changes made to the latter are also made to the former.

```
def copylist(L):
    import copy
    return copy.deepcopy(L)
```

2.9 Printing Functions

The function print_list(A,direction) prints the inside of a list either horizontally or vertically depending on the value of the parameter direction, which takes the values horizontal and vertical.

```
---------------------------------------------------------------
def print_list(A,direction):      # direction horizontal or vertical
    for item in A:
        if direction == 'horizontal':
            print(item,end = ' ')              # print horizontally
        if direction == 'vertical':
            print(item)                         # print vertically

-------------------------- Sample Run --------------------------
Input:
A = ['1', '2', '3']
print(A,'\n')
print_list(A,'horizontal'); print('\n')
print_list(A,'vertical')
---------------------------------------------------------------
Output:
['1', '2', '3']

1 2 3

1
2
3
---------------------------------------------------------------
```

The function format_print(A,nspaces,flush) takes a double list and prints a formatted table version with the columns flush left or flush right, depending on the value of the string parameter flush (='left','center','right'). The parameter nspaces add spaces to the columns. The spaces between adjacent columns vary depending on the width of the entries.

```
---------------------------------------------------------------
def format_print(A, nspaces, flush):
    if len(A) == 0: return
    if isinstance(A,str):                           # if string,
        print(A); return                            # return it
    if isinstance(A,list):                          # if list,
        if not isinstance(A[0],list):        # but not double list,
            print_list(A,'horizontal')       # print horizontally
            return
    nrows, ncols = len(A), len(A[0])                # dimensions of A
    col_width = [0 for j in range(len(A[0]))]
    B = copylist(A)                                 # don't destroy A
```

```
        for i in range(nrows):
            for j in range(ncols):              # get width of each column
                if len(A[i][j]) > col_width[j]: # col width[j]=width of
                    col_width[j] = len(A[i][j]) # largest entry in col j
        for i in range(nrows):                  #   add spaces to A's entries
            for j in range(ncols):
                # spaces to left or right of entry:
                this_many_spaces = col_width[j] + nspaces - len(A[i][j])
                width = ' '*this_many_spaces
                half_width = ' '*((this_many_spaces+1)//2)
                if flush == 'left':
                    B[i][j] = B[i][j] + width
                elif flush == 'right':
                    B[i][j] = width + B[i][j]
                else:                                   # entry in middle
                    B[i][j] = half_width + B[i][j] + half_width
        for i in range(len(B)):
            for entry in B[i]:
                print(entry, end = '')              # print on single line
            if i < len(B) - 1:
                print('')

--------------------------- Sample Run ---------------------------
Input:
s = '1, 2+3i, 1.2356789,4; 5, 6, 7, 8+9i'
A = string2table(s)
print('left flush:')
format_print(A,2,'left')
print('\n')

print('center:')
format_print(A,2,'center')
print('\n')

print('right flush:')
format_print(A,2,'right')
------------------------------------------------------------------
Output:
left flush:
1  2+3i  1.2356789  4
5  6     7          8+9i

center:
1  2+3i  1.2356789    4
5   6        7       8+9i

right flush:
1  2+3i  1.2356789     4
5     6          7  8+9i
------------------------------------------------------------------
```

2.9 Printing Functions

The function `print_fraction(prepend,num,den,append)` uses the preceding function to print the fraction num/den in a large format. The arguments `prepend` and `append` are optional strings.

```
def print_fraction(prepend,num,den,append):
    maxlen = max(len(num),len(den))
    leftspace = ' '*len(prepend)
    dash = '-'*maxlen
    A = [[leftspace+num],[prepend+ dash + append],[leftspace+den]]
    format_print(A, 1, 'center')

--------------------------- Sample Run ---------------------------
Input:
print_fraction('  ','2468 ','abcdefg','=')
print('\n')
print_fraction('2*','1234 ','abcdefg','')
------------------------------------------------------------------
Output:

  2468
  ------- =
  abcdefg

    1234
  2*-------
    abcdefg
------------------------------------------------------------------
```

Symbolic Logic

3

We have already had some informal exposure to logic through the Python operators and,or,not. In this section we formalize the notions underlying these operators and implement in Python a propositional (Boolean) algebra that is similar in many ways to ordinary algebra. The functions in the chapter comprise the module Logic.py. Here is the header for the module.

```
-------------------------- Logic.py --------------------------
import Tools as tl
```

3.1 Compound Statements

A *statement* or *proposition* is a declarative sentence which is either true or false but not both. We shall use the letters $p, q, r, \ldots$ to designate so-called *simple statements*. These are given the *truth values* 1 for true and 0 for false and so may be taken as variables.

Compound statements are constructed from simple statements p, q using the logical operations

(a) *conjunction* pq (p and q).
(b) *disjunction* $p + q$ (p or q).
(c) *negation* p' (not p).
(d) *conditinal* $p \to q$ (p implies q, if p then q).
(e) *biconditional* $p \leftrightarrow q$ (p if and only if q).

Fig. 3.1 Truth table for the basic logical operations

p	q	pq	$p+q$	p'	$p \to q$	$p \leftrightarrow q$
1	1	1	1	0	1	1
1	0	0	1	0	0	0
0	1	0	1	1	1	0
0	0	0	0	1	1	1

The notation for these operations is chosen for ease in typing compound statements. The truth values of these statements are displayed by the *truth table* in Fig. 3.1. The first two columns list in conventional format all possible truth values 0,1 of the pair of simple statements p, q. The remaining columns give the corresponding truth values of the above compound statements.

Note that the conditional $p \to q$ is true even if p is false. This is called the *principle of explosion* ("from falsehood anything follows"). Note also that the biconditional $p \leftrightarrow q$ is true precisely when p and q have the same truth values.

Like ordinary algebra, propositional algebra has a *hierarchy* or *precedence* of operations. For example, the negations in the statement $pq' + p'r$ are evaluated first, then the conjunctions, and finally the disjunction. A similar hierarchy is observed in the statement $pq' \to p'r$, the conditional being evaluated last. For more complex statements parentheses are needed. For example, in the statement $(p + p'r) \leftrightarrow (pq' \to r)$ the parentheses force the biconditional to be evaluated last; removing the parentheses would result in a different statement. The module developed in this chapter conforms to these precedence rules.

3.2 Generating a Truth Table

The main function of the module, `statement2truthtable(stmt)`, takes a compound statement and returns its truth table in list form. It also returns two other tables, one that contains only those rows for which the statement is true, the other containing only those rows for which the statement is false. A companion function, `print_truth_table(table)` prints the tables. Here is a sample run:

```
-----------------------------------------------------------------
Input:
stmt = "((p+q+r)->qr')<->(pq)"               # input statement
tableA,tableT,tableF = statement2truthtable(stmt)
print_truth_table(tableA); print('\n')       # complete table
print_truth_table(tableT)  print('\n')       # true rows only
print_truth_table(tableF)                    # false rows only
-----------------------------------------------------------------

Output:
pqr  ((p+q+r)->qr')<->(pq)    # the statement is part of the table
111    0
110    1
101    1
```

3.2 Generating a Truth Table

```
100  1
011  1
010  0
001  1
000  0

pqr ((p+q+r)->qr')<->(pq)
110  1
101  1
100  1
011  1
001  1

pqr ((p+q+r)->qr')<->(pq)
111  0
010  0
000  0
```

The code for the function includes three print statements, allowing the user to follow the progression of the program. These may be commented out.

```
def statement2truthtable(stmt):
    global idx
    stmt = stmt.replace(" ","")              # remove white space
    var_list, varstring = tl.get_vars(stmt)  # variables in stmt
    itab = initial_table(len(varstring))     # variable truth values
    tableT = []             # table with only true values attached
    tableT.append(varstring + ' ' + stmt)    # attachment stmt
    tableF = []             # table with only false values attached
    tableF.append(varstring + ' ' + stmt)
    tableA = []             # table with all values attached
    tableA.append(varstring + ' ' + stmt)
    stmt = tl.insert_asterisks(stmt,var_list)
    for i in range(len(itab)):          # generate the truth values
        row = itab[i]
        ps = stmt                       # for populating with 0"s, 1"s
        print("i:",i,', ',end = " ")            # for observation
        for j in range(len(row)):       # insert 1"s, 0"s into ps
            ps = ps.replace(varstring[j],row[j])
            print(ps,end = " ")                 # for observation
        idx = 0                  # points to position in string ps
        value = eval_stmt(ps,0)  # truth value of populated stmt
        #print(value,"\n")                       # for observation
        row = row + value        # attach truth value of statement
        tableA.append(row)                       # all rows
        if value == '1': tableT.append(row)      # only true rows
        if value == '0': tableF.append(row)      # only false rows
```

The function `insert_asterisks` in the above code is used for easy processing of conjunctions. For example, applying the function to ``(p->r'q)(q->(r+p'))`` yields ``(p->r'*q)*(q->(r+p'))``. Double quotes are needed here so that single quotes will be processed by Python as intended, namely as negation symbols and not as string quotes. The function `print_truth_table` removes the quotes in the table for clarity. The function `get_vars` retrieves the variable names in the statement. The function `initial_table` makes columns of 1's and 0's for the variables. The function `eval_stmt` takes a statement populated with 1's and 0's and returns its truth value, which is then attached to the end of the current row of the table. Here is the code for the print function:

```
def print_truth_table(table):
    print(table[0].replace('"',''))         # remove double quotes "
    s = len(table[1])*' '                   # space to separate stmt truth value
    p = len(table[1])-1                     # put in space starting here
    for i in range(1,len(table)):
        row = tl.insert_string(table[i],s,p)    # insert space
        print(row.replace("'",''))              # remove single quotes
```

The function `initial_table(numvars)` takes as input the number of variables in a statement and generates the initial columns of a truth table. The columns consist of all possible combinations of zeros and ones in standard format.

```
def initial_table(numvars):
# generates a table of 1"s,0"s in standard format
    table = []
    ncols = numvars
    nrows = 2**numvars                      # number of rows of zeros and ones
    for i in range(1,nrows+1):              # generate the rows
        row = ''
        for j in range(1,ncols+1):
            tval = zero_one(i,j,ncols)      # truth val at (i,j)
            row = row + str(tval)           # append to row
        table.append(row)
    return table

--------------------------- Sample Run ---------------------------
Input:
tl.print_list(initial_table(4),'h')         # print horizontally
------------------------------------------------------------------
Output:
1111 1110 1101 1100 1011 1010 1001 1000 0111 0110 0101 0100 0011 \
0010 0001 0000
```

3.2 Generating a Truth Table

The function `zero_one(i,j,n)` creates all possible zero-one values of n variables $p, q, \ldots$. It is based on the formula

$$(i-1)/2^{n-j}, \quad 1 \le i \le 2^n, \quad 1 \le j \le n,$$

where i is a row number, and j is a column number. For the case of three variables the formula generates the following fractions in tabular form with rows $i = 1, 2, \ldots, 8$ and columns $j = 1, 2, 3$.

$$\begin{bmatrix} 0/4 & 0/2 & 0/1 \\ 1/4 & 1/2 & 1/1 \\ 2/4 & 2/2 & 2/1 \\ 3/4 & 3/2 & 3/1 \\ 4/4 & 4/2 & 4/1 \\ 5/4 & 5/2 & 5/1 \\ 6/4 & 6/2 & 6/1 \\ 7/4 & 7/2 & 7/1 \end{bmatrix}$$

Taking integer part of these fractions yields the table of x-values

$$\begin{bmatrix} 0 & 0 & 0 \\ 0 & 0 & 1 \\ 0 & 1 & 2 \\ 0 & 1 & 3 \\ 1 & 2 & 4 \\ 1 & 2 & 5 \\ 1 & 3 & 6 \\ 1 & 3 & 7 \end{bmatrix}$$

Applying the transformation `y = int(1 + (-1)^x)/2` to these values produces the desired truth table

$$\begin{bmatrix} 1 & 1 & 1 \\ 1 & 1 & 0 \\ 1 & 0 & 1 \\ 1 & 0 & 0 \\ 0 & 1 & 1 \\ 0 & 1 & 0 \\ 0 & 0 & 1 \\ 0 & 0 & 0 \end{bmatrix}$$

Here is the code that produces the entries of the above table.

```
def zero_one(i,j,ncols):   # returns truth value at row i and col j
    x = int((i-1)/2**(ncols-j))    # 1 <= i <= nrows, 1 <= j <= ncols
    y = int((1 + (-1)**x) / 2)
    return y
```

3.3   The Calculation Engine

The calculations in the module are performed by the function eval_stmt(ps,mode), which takes a statement ps populated with 0's and 1's and returns its truth value. The variable mode determines the order in which the calculations are performed. A global variable idx points to the characters in ps; it is set to 0, the index of the first character of ps, for each populated statement. Here is the code:

```
def eval_stmt(ps,mode):
    ps = ps.replace("1'","0")           # deal with negation first
    ps = ps.replace("0'","1")
    global idx
    while idx < len(ps):
        c = ps[idx]                      # character at index idx
        if c in "01":
            p = c                        # c is a truth value
            idx += 1
        elif c == "+":
            if mode > 0: break           # wait for higher mode to finish
            idx += 1                                 # go to lowest mode...
            q = eval_stmt(ps,0)          # all other calculations come 1st
            p = disjunction(p,q)                     # calculate p+q
        elif c == "*":
            idx += 1
            q = eval_stmt(ps,2)          # highest mode:conjunctions first
            p = conjunction(p,q)                     # calculate p*q
        elif c == "-":                                       # conditional
            if mode > 1: break           # wait for conjunction calculation
            idx += 2                                         # skip "->"
            q = eval_stmt(ps,1)
            p = conditional(p,q)                     # calculate p->q
        elif c == "<":                                       # biconditional
            if mode > 1: break
            idx += 3                                         # skip "<->"
            q = eval_stmt(ps,1)
            p = biconditional(p,q)
        elif c == "(":
            idx += 1                                         # skip "("
            p = eval_stmt(ps,0)          # evaluate stuff inside ()
            idx = idx + 1                                    # skip ")"
```

3.3 The Calculation Engine

```
            if idx < len(ps) and ps[idx] == "'":
                p = negation(p)         # negate paren expression
                idx += 1
        elif c == ")":
            break
    return p
```

To see how the function works, consider the populated statement ps=1*0->1. In the initial iteration of the while loop (idx = 0), the function reads the first 1 in the statement, sets the variable *p* to 1, increments the index, and performs the next iteration. At this point ps[idx] = '*', so the case c = '*' is activated. The function then increments idx and calls itself to extract the next value 0 of ps. While still within the called version of the function, the next iteration is performed. At this point ps[idx] = '-' (the first part of the conditional symbol ->) hence the case c = '-' is activated. However, because the current mode is bigger than 1, the while loop breaks, the value 0 is returned by the called version of the function, and the conjunction is performed within the original version of the function. The value of the mode in this version is 0 so, since we still have ps[idx] = '-', the conditional may finally be performed. In this way we have ensured that the conjunction is performed before the conditional, so the desired order (1*0)->1 is achieved rather than the order 1*(0->1). A similar analysis may be carried out on the populated statement 1*0 + 1, interpreted by the function as (1*0)+1. Here, the conjunction is evaluated first, using recursion, and the disjunction is evaluated last.

Here is the code for the logical operations. The reader may check these by substituting the values 0,1.

```
def disjunction(p, q):
    return str(min(int(p) + int(q), 1))

def conjunction(p, q):
    return str(int(p)*int(q))

def negation(p):
    return str(1-int(p))

def conditional(p,q):
    return disjunction(negation(p), q)

def biconditional(p,q):
    return conjunction(conditional(p,q), conditional(q,p))
```

The following scheme shows the evaluation the statement ''(p+q)->pqr'' in the (i, j) for loops of statement2truthtable(stmt).

	j = 0	j = 1	j = 2	value
i = 1:	(1+q)->1qr'	(1+1)->11r'	(1+1)->111'	0
i = 2:	(1+q)->1qr'	(1+1)->11r'	(1+1)->110'	1
i = 3:	(1+q)->1qr'	(1+0)->10r'	(1+0)->101'	0
i = 4:	(1+q)->1qr'	(1+0)->10r'	(1+0)->100'	0
i = 5:	(0+q)->0qr'	(0+1)->01r'	(0+1)->011'	0
i = 6:	(0+q)->0qr'	(0+1)->01r'	(0+1)->010'	0
i = 7:	(0+q)->0qr'	(0+0)->00r'	(0+0)->001'	1
i = 8:	(0+q)->0qr'	(0+0)->00r'	(0+0)->000'	1

3.4 Equivalent Statements

A compound statement that is true for all values of its variables is called a *tautology*. For example, the statement p+p' is a tautology; its value is always 1. A compound statement that is false for all values of its variables is called a *contradiction*. An example is pp'; its value is always 0. The negation of a tautology is a contradiction and vice-versa. The following functions check for tautologies and contradictions. It uses the output of the function statement2truthtable, namely, Atable, Ttable, and Ftable (giving, respectively, all truth values, only true values, and only false values) and compares the lengths of the tables. For example, if len(Ttable)=len(Atable) then the statement is a tautology.

```
-----------------------------------------------------------------
def is_tautology(stmt):
    # returns True if Atable and Ttable are same size.
    Atable,Ttable,Ftable = statement2truthtable(stmt)
    return len(Ttable) == len(Atable)
                                            # True if tautology
def is_contradiction(stmt):
    # returns True if Atable and Ftable are same size.
    Atable,Ttable,Ftable = statement2truthtable(stmt)
    return len(Ftable) == len(Atable)       # True if contradiction

-------------------------- Sample Run --------------------------
Input:
stmt = '(p->q)(q->r)->(p->r)'
print(is_tautology(stmt,varstring))
-----------------------------------------------------------------
Output:
True
-----------------------------------------------------------------
```

3.5 Valid Arguments

Two compound statements a and b are said to be *equivalent*, written $a \equiv b$, if one statement is true is whenever the other is true, that is, the statements have precisely the same truth tables. For example, the statements $p' + q$ and $p \rightarrow q$ have identical truth tables, hence are equivalent.

Another way to determine the equivalence of statements a and b is to check if the biconditional $a \leftrightarrow b$ is always true, that is, if it is a tautology. The following function uses this idea to test for equivalence. It takes a pair of statements a, b and applies the function is_tautology(stmt) to the statement $a \leftrightarrow b$.

```
def are_equivalent(a,b):
    stmt = "(" + a + ")<->(" + b + ")"        # form biconditional
    return is_tautology(stmt)

------------------------------ Sample Run ------------------------------
Input:
print(are_equivalent("(p + q)'" ,"p'q'"))
print(are_equivalent("(pq)'"    ,"p' + q'"))
print(are_equivalent("p -> q"   ,"q' -> p'"))
print(are_equivalent("p(q + r)" ,"pq + pr"))
print(are_equivalent("p + qr","(p + q)(p + r)"))
print(are_equivalent("(p+q)+r"  ,"p+(q+r)"))
print(are_equivalent("(pq)r"    ,"p(qr)"))

Output:
True            # DeMorgan's law
True            # DeMorgan's law
True            # contrapositive
True            # distributive law
True            # distributive law
True            # associative law
True            # associative law
```

3.5 Valid Arguments

An *argument* in logic is a sequence of statements $a_1, a_2, \ldots, a_n$, called *premises*, together with statement b, called the *conclusion*. An argument is said to be *valid* if b is true whenever all the statements $a_1, a_2, \ldots, a_n$ are true, that is, whenever the conjunction $a_1 a_2 \cdots a_n$ is true. Figure 3.2 gives some well-known valid arguments displayed in standard form: the premises placed above a line and the conclusion below.

hypothetical syllogism	disjunctive syllogism	modus ponens	modus tollens
$p \to q$	$p + q$	$p \to q$	$p \to q$
$q \to r$	p'	p	q'
$p \to r$	q	q	p'

Fig. 3.2 Valid arguments

The following program takes a list argument of premises and a conclusion (the last entry) returns True if the argument is valid and False otherwise.

```
def isvalid(argument):
    argument = argument.replace(" ","")
    premises, conclusion = argument.split(";")
    premises = premises.split(",")
    for i in range(len(premises)):
        premises[i] = "(" + premises[i] + ")"
    premise = "".join(premises)
    stmt = premise + "->(" + conclusion + ")"
    if is_tautology(stmt): return "valid"
    return "not valid"
```
--------------------------- Sample Run ---------------------------
```
Input:
arg1 = ["p->q","q->r","p->r"]
arg2 = ["p+q","p'","q"]
arg3 = ["p->q","p","q"]
arg4 = ["p->q","q'","p'"]
arg5 = ["p->q","q","p"]
arg6 = ["p+q","q","p"]
arg7 = ["p->q","p'","q"]
arg8 = ["p'->q","pq","q'"]
args = [arg1,arg2,arg3,arg4,arg5,arg6,arg7,arg8]

for i in range(len(args)):
print('arg',i+1,isvalid(args[i]))
```

```
Output:
arg 1 valid           # hypothetical syllogism
arg 2 valid           # disjunctive syllogism
arg 3 valid           # modus ponens
arg 4 valid           # modus tollens (contrapositive)
arg 5 not valid
arg 6 not valid
arg 7 not valid
arg 8 not valid
```

3.6 Disjunctive Normal Form

In this section and the next we construct functions that are the reverse of the function `statement2truthtable(stmt)` in that they take as input a set of truth values and output a statement with those values. This is useful in the design logic circuits in computer science, which are transistor circuits that have input and output of values 1 (high voltage) and 0 (no voltage).

The first such function `tvals2conj` takes a string `valcombo` of values and a string `varstring` of variables and produces a conjunction that is true for precisely these values. For example if `varstring` = ``pqr'' and `valcombo` = ``101'', then the function returns `varstring` = ``pq'r''.

```
def tvals2conj(varstring,valcombo):
# true precisely for the combination of values in valcombo
    conj = ''
    for i in range(len(varstring)):
        conj = conj + varstring[i]          # attach the variable
        if valcombo[i] == '0':              # false value?
            conj = conj + "'"      # attach additionally a negation
    return conj
```

```
---------------------------- Sample Run ----------------------------
Input:
varstring = "pqrstuvwxy"
valcombo  = "1010101010"
print(tvals2conj(varstring,valcombo))
--------------------------------------------------------------------
Output:
pq'rs'tu'vw'xy'    # true exactly for values in valcombo "1010101010"
--------------------------------------------------------------------
```

The function `disj_of_conj` takes a string of variables and a string of comma-separated zero-one combinations and returns a statement that is true for each of these sequences. It does so by forming the disjunction of the conjunctions that are true for each of the zero-one combinations.

```
--------------------------------------------------------------------
def disj_of_conj(varstring,valcombos):
    disj = ''
    valcombos = valcombos.split(',')
    for combo in valcombos:              # run through desired combos
        conj = tvals2conj(varstring,combo)
        disj = disj + ' + ' + conj  # form the disjunction; pad the +
    return disj[3:]                      # remove first padded +
```

```
---------------------------- Sample Run ----------------------------
Input:
varstring = "pqr"
```

```
valcombos = "100,111,101"
print(disj_of_conj(varstring,valcombos))
```
```
Output:
pq'r' + pqr + pq'r     # true for each valcombo; false for all others
```

In the sample run, the statement ``pq'r''' is true precisely for the combo ``100'', ``pqr'' is true precisely for the combo ``111'', and ``pq'r'' is true precisely for the combo ``101''. Thus the disjunction ``pq'r'+pqr+pq'r'' is true for each of these combos and no others. A statement of this form is called a *disjunction of basic conjunctions*.

The following function takes any statement and returns an equivalent statement which is a disjunction of basic conjunctions, called the *disjunctive normal form* of the statement. It takes the rows of Ttable without the header, joins them into a comma-separated string of value combinations and feeds the string to disj_of_conj. Print statements are included to illustrate the program's progress.

```
def disjunctive_form(stmt):
    stmt = stmt.replace(" ","")                  # remove white space
    varstring = tl.get_vars(stmt)[1]             # string form
    Ttable = statement2truthtable(stmt)[1][1:]   # true rows only
    print('Ttable',Ttable)
    truecombos = ','.join(Ttable)                # convert table to string
    print('truecombos',truecombos)               # for observation
    return disj_of_conj(varstring,truecombos)
```
```
-------------------------- Sample Run --------------------------
Input:
stmt = "((p+q+r)->qr')<->(pq)"
disj = disjunctive_form(stmt)
print('disunctive form:',disj)
```
```
Output:
Ttable:     ['1101', '1011', '1001', '0111', '0011']
truecombos: 1101,1011,1001,0111,0011
disunctive form: pqr' + pq'r + pq'r' + p'qr + p'q'r
```

3.7 Conjunctive Normal Form

The function tvals2disj(varstring,valcombo) is the analog of the function tvals2conj(varstring,valcombo). It takes a string varstring of variables and a string valcombo of ones and zeros and produces a disjunction that is false for precisely these val-

3.7 Conjunctive Normal Form

ues. For example if `varstring` = `''pqr''` and `valcombo` = `''110''`, then the function returns `''p'+q'+r''`

```
def tvals2disj(varstring,valcombo):
    #false precisely for valcombo
    var_list = list(varstring)
    valcombo = list(valcombo)
    for i in range(len(var_list)):
        if valcombo[i] == '1':
            var_list[i] = var_list[i] + "'" + "     # negate
        else:
            var_list[i] = var_list[i] + " " + "
    disj = ''.join(var_list)
    return disj[:len(disj)-3]       # remove last padded plus sign

---------------------------- Sample Run ----------------------------
Input:
varstring = "pqrstuvwxy"
valcombo  = "1010101010"
print(tvals2disj(varstring,valcombo))
---------------------------------------------------------------------
Output:
p'+q+r'+s+t'+u+v'+w+x'+y     # false exactly for combo "1010101010"
---------------------------------------------------------------------
```

The function `conj_of_disj(varstring,valcombos)` takes a string of variables and a string of value combinations and returns a statement that is false for each of these combinations and no others. It does so by forming the conjunction of the disjunctions that are false for the value combos.

```
def conj_of_disj(varstring,valcombos):
    conj = ''
    valcombos = valcombos.split(',')
    for combo in valcombos:            # run through desired combos
        disj = tvals2disj(varstring,combo) #
        conj = conj + '(' + disj + ')'     # form the disjunction
    return conj

---------------------------- Sample Run ----------------------------
Input:
varstring = "pqr"
valcombos = "100,111,101"
print(conj_of_disj(varstring,valcombos))
---------------------------------------------------------------------
Output:
(p' + q + r)(p' + q' + r')(p' + q + r')
---------------------------------------------------------------------
```

In the sample run the statement ``(p'+q+r)`` is false precisely for the combination ``100``, ``(p'+q'+r')`` is false precisely for the combination ``111``, and ``(p'+q+r')`` is false precisely for the combination ``101``. Thus the conjunction ``(p'+q+r)(p'+q'+r')(p'+q+r')`` is false for each of these combinations and no others. A statement of the form ``(p'+q+r)(p'+q'+r')(p'+q+r')`` is called a *conjunction of basic disjunctions*.

The following function takes a statement and returns an equivalent statement which is a conjunction of basic disjunctions, called the *conjunctive normal form* of the statement.

```
def conjunctive_form(stmt):
    stmt = stmt.replace(" ","")              # remove white space
    varstring = tl.get_vars(stmt)[1]         # string form
    Ftable = statement2truthtable(stmt)[2][1:]  # false rows only
    false_combos = ','.join(Ftable)
    return conj_of_disj(varstring,false_combos)
```

------------------------- Sample Run -------------------------
Input:
stmt = "((p+q+r)->qr')<->(pq)"
print(conjunctive_form(stmt))

Output:
(p' + q' + r')(p + q' + r)(p + q + r)

Properties of Integers

4

In this chapter we explore properties of integers, with particular attention to prime numbers. The functions in the chapter comprise the module Logic.py. The module is headed by

```
------------------------- Number.py -------------------------
import math as ma                              # Python package
import Tools as tl
-------------------------------------------------------------
```

4.1 Number Bases

Recall that an integer in base 10 is represented by a sequence of digits which are multiplied by suitable powers of 10. For example,

$$1234567 = \underline{1} \cdot 10^6 + \underline{2} \cdot 10^5 + \underline{3} \cdot 10^4 + \underline{4} \cdot 10^3 + \underline{5} \cdot 10^2 + \underline{6} \cdot 10^1 + \underline{7} \cdot 10^0.$$

We have underlined the digits to highlight their role in the expansion. The position of a digit determines the power of ten. There is nothing particularly special about the number 10, however. Indeed, any integer greater than 1 may serve as a base. For example, in base 5 the same number may be expressed as

$$304001232_5 = \underline{3} \cdot 5^8 + \underline{2} \cdot 5^6 + \underline{1} \cdot 5^3 + \underline{2} \cdot 5^2 + \underline{3} \cdot 5^1 + \underline{2} \cdot 5^0.$$

Here we have used a subscript to designate the base, omitted for base 10. For base 5 the digits 0–4 are available. For base 2, the so called "binary system," only the digits 0 and 1 are allowed.

This makes binary numbers well-suited for computer calculations, since standard computers use two-state technology. The number 1234567 in base 2 is 100101101011010000111.

To convert a number n from base 10 to base b one repeatedly divides by b; the remainders are the desired base b digits. To illustrate, consider the base 7 expansion

$$n = d_4 \cdot 7^4 + d_3 \cdot 7^3 + d_2 \cdot 7^2 + d_1 \cdot 7^1 + d_0 \cdot 7^0.$$

Dividing n by 7 yields

$$n = q_0 \cdot 7 + d_0, \text{ where } q_0 = d_4 \cdot 7^3 + d_3 \cdot 7^2 + d_2 \cdot 7^1 + d_1.$$

The number q_0 in the division is the quotient and the digit d_0 is the remainder, a number between 0 and $b - 1$ inclusive. Dividing q_0 by 7 produces the next digit d_1:

$$q_0 = q_1 \cdot 7 + d_1, \text{ where } q_1 = d_4 \cdot 7^2 + d_3 \cdot 7^1 + d_2.$$

The process continues until all the digits have been generated.

Capital letters are typically used for bases 10–36. Of particular importance in computer science is the base 16 or hexadecimal system, which uses the digits 0–9 and the letters A–F. Here A represents the number 10, B the number 11, etc. The number 1234567 in base 16 is

$$12D687_{16} = 1 \cdot 16^5 + 2 \cdot 16^4 + 13 \cdot 16^3 + 6 \cdot 16^2 + 8 \cdot 16^1 + 7 \cdot 16^0.$$

For bases larger than 36 one can use lower case letters as well, but this too is limited. Alternatively, one can use a list for the digits. For example, the base 10 number 1234567 in base 99, which has the expansion

$$\underline{1} \cdot 99^3 + \underline{26} \cdot 99^2 + \underline{95} \cdot 99^1 + \underline{37} \cdot 99^0,$$

could be represented by the digit list [1,26,95,37], where it is acknowledged somewhere that the base is 99, say by attaching it to the beginning of the list.

The functions in this section incorporate both ideas. The first of these, value_to_digits(n,b), takes a positive integer n, which we shall refer to throughout the section as a *value*, and a base b, and returns the digit list together with the base as the first member of the list. The function uses the floor division operator % and modulo operator // discussed in Chap. 1.

```
def value_to_digits(n, b):
    # returns the base b digit list of the value n
    if n == 0:
        return [0]
    base_digit_list = []
    while n > 0:
        digit = n % b                         # digit is remainder
        base_digit_list = [digit]+ base_digit_list   # attach digit
```

4.1 Number Bases

```
            n = n // b                          # divide out the b
            base_digit_list = [b] + base_digit_list   # attach base to list
            return base_digit_list

-------------------------- Sample Run --------------------------
n = 12345678901234567890
b = 2025
print(value_to_digits(n,b))
----------------------------------------------------------------
Output:
[2025, 362, 1150, 1666, 1308, 268, 1440]       # b the first entry
----------------------------------------------------------------
```

For an explicit expansion in terms of powers of the base one can use the following function:

```
----------------------------------------------------------------
def digits_to_expansion(base_digit_list):
    # converts a base b digit list to a base b expansion
    expansion = ''
    b = base_digit_list[0]
    digits = base_digit_list[1:]               # extract base
    L = len(digits)
    for i in range(L):
        exp = str(L-i-1)
        digit = digits[i]                      # get a digit from list
        if digit == 0: continue
        # attach current digit*b^power to expansion:
        if exp == '0':
            expansion = expansion+ ' + ' +str(digit)
        elif exp == '1':
            expansion = expansion+ ' + ' +str(digit)+ '*' +str(b)
        else:
            expansion = expansion+ ' + ' +str(digit)+ '*' + \
                str(b) + '^' + exp
    return expansion[3:]                       # remove last padded '+'

-------------------------- Sample Run --------------------------
Input:
[2025, 362, 1150, 1666, 1308, 268, 1440]       # base = 2025
----------------------------------------------------------------
Output:
362*2025^5 + 1150*2025^4 + 1666*2025^3 + 1308*2025^2 + 268*2025^1\
 + 1440*2025^0
----------------------------------------------------------------
```

The function digits_to_value is the reverse of the function value_to_digits. It takes a list of digits and a base and returns a value. Note that the digit values must be smaller than the base, since the former are remainders upon division by the latter. The function uses the Python function eval() to get the value.

```
def digits_to_value(base_digit_list):
    ex = digits_to_expansion(base_digit_list)
    ex = ex.replace('^','**')      # eval uses ** for exponentiation
    return eval(expansion)
```
---------------------------- Sample Run ----------------------------
```
Input:
b = 2025
base_digit_list = [2025,362, 1150, 1666, 1308, 268, 1440]
value = digits_to_value(base_digit_list)
print(value)
```

```
Output:
12345678901234567890
```

The function value_to_symbols(n,b) below is a string version of the function value_to_digi It converts a positive integer (value) n into a string of digit symbols. This works only for $b \leq 62$, the number of available symbols.

```
def value_to_symbols(n,b):  # converts value n into a symbol string
    if b > 62: return              # not enough digits available
    digits = value_to_digits(n,b)                # get digit list
    digit_symbols = '0123456789ABCDEFGHIJKLMNOPQRSTUVWXYZ' \
                    + 'abcdefghijklmnopqrstuvwxyz'
    symbolstring = ''
    for digit in digits:           # replace each digit by its symbol
        symbolstring = symbolstring + digit_symbols[digit]
    return symbolstring
```
---------------------------- Sample Run ----------------------------
```
Input:
n = 12345678901234567890
print(value_to_symbols(n,2))
print(value_to_symbols(n,8))
print(value_to_symbols(n,16))
print(value_to_symbols(n,62))
```

```
Output:
1010101101010100101010011000110011101011000111110000101011010010
1255245230635307605322
AB54A98CEB1F0AD2
EhzL6HwZ5ow
```

The function symbols_to_value is the reverse of value_to_symbols(n,b). It converts a base $b \leq 62$ symbol string into a positive integer (value). The function uses the index

4.1 Number Bases

method to find the position of the symbol in the `digit_symbols` list, the position being the numerical value of the digit.

```
def symbols_to_value(symbolstring,b):
    # converts a symbol string in base b to its numerical value
    digit_symbols = '0123456789ABCDEFGHIJKLMNOPQRSTUVWXYZ' \
                    + 'abcdefghijklmnopqrstuvwxyz'
    digit_list = []
    for symbol in symbolstring:
        index = digit_symbols.index(symbol)    # position of symbol
        digit_list.append(index)
    return digits_to_value([b]+digit_list)     # append base to list

-------------------------- Sample Run --------------------------
Input:
print(symbols_to_value('AB54A98CEB1F0AD2',16))

Output:
12345678901234567890
```

The last function in the section is a combination of the preceding two. It converts a base a digit string into a base b digit string. The example suggests a possible use in transmitting secure messages. Here, both the sender and receiver are in possession of the key (a, b), a pair of numbers that "locks" and "unlocks" the message. Typically the message is written in digits and uppercase letters, whose values are ≤ 36. The bases a and b are then chosen to larger that 36. The sample run conforms to these restrictions.

```
def base2base(symbolstring,a,b):
    n = symbols_to_value(symbolstring,a)
    return value_to_symbols(n,b)

-------------------------- Sample Run --------------------------
Input:
a = 52; b = 62
message = 'THEFILESMUSTBEDELETED'
print(message)
coded_message = base2base(message,a,b)
print(coded_message)
decoded_message = base2base(coded_message,b,a)
print(decoded_message)

Output:
THEFILESMUSTBEDELETED                          # message
rwjLPPTCo27P9yZD4ZHx                           # encoded message
THEFILESMUSTBEDELETED                          # decoded by recipient
```

4.2 Divisibility

An integer b is said to *divide* an integer a if there exists an integer q such that $a = qb$. The integers b and q are called *divisors* or *factors* of a. For example, $\pm a$ and ± 1 are always divisors of a. Positive divisors of $a > 0$ other than a and 1, are called *proper*. Note that a proper divisor b of $a = qb$ must satisfy the inequality $b \leq a/2$. Indeed, if the reverse inequality held we would have $a < 2b \leq qb = a$. If b divides a we write $b \mid a$; otherwise we write $b \nmid a$. Thus $3 \mid 15$ but $4 \nmid 15$.

A positive integer with no proper divisors is said to be *prime*; otherwise, it is said to be *composite*. Prime numbers figure prominently in pure mathematics and have important applications in cryptography.

Here's a simple program that grinds out the proper divisors of a number $a > 0$ by checking if $a \% b = 0$ for all $b \leq a/2$. This, of course, is inefficient for large numbers on slow computers.

```
def generate_divisors(a):
    divisors = []
    for b in range(2,int(a/2)+1):
        if a%b == 0:                                    # b | a ?
            divisors.append(b)                          # yes, attach divisor
    divisors = list(set(divisors))                      # eliminate duplicates
    return sorted(divisors)                             # sorted for clarity

-------------------------- Sample Run --------------------------
Input:
print(generate_divisors(2**5*3**6))

Output:
[2, 3, 4, 6, 8, 9, 12, 16, 18, 24, 27, 32, 36, 48, 54, 72, 81, 96,
108, 144, 162, 216, 243, 288, 324, 432, 486, 648, 729, 864, 972,
1296, 1458, 1944, 2592, 2916, 3888, 5832, 7776, 11664]
```

4.3 Extended Euclidean Algorithm

The *greatest common divisor* or *gcd* of a pair of integers a, b is a positive integer g with the properties

- g divides a and b,
- if d divides a and b then d divides g.

The gcd of a and b is denoted by $\gcd(a, b)$. We also set

4.3 Extended Euclidean Algorithm

$$\gcd(-a, b) = \gcd(a, -b) = \gcd(-a, -b) = \gcd(a, b).$$

The *standard Euclidean algorithm* is a way of computing the gcd. The algorithm takes a pair of positive integers $a > b$ and computes a sequence of quotients $q_1, q_2, \ldots, q_n$ and a sequence of remainders $r_0, r_1, \ldots, r_{n+1}$ satisfying the following equalities:

$$r_0 = a \quad r_1 = b$$
$$r_2 = r_0 - q_1 r_1, \quad 0 < r_2 < r_1$$
$$\vdots$$
$$r_{k+1} = r_{k-1} - q_k r_k, \quad 0 < r_{k+1} < r_k \qquad (4.1)$$
$$\vdots$$
$$r_{n-2} = r_n - q_{n-1} r_{n-1}, \quad 0 < r_n < r_{n-1}$$
$$r_{n-1} = r_{n+1} - q_n r_n, \quad 0 = r_{n+1} < r_n.$$

The equation $r_{k+1} = r_{k-1} - q_k r_k$ may be rewritten as $r_{k-1} = q_k r_k + r_{k+1}$, which is the division algorithm applied the pair (r_{k-1}, r_k), thus generating the new pair (q_k, r_{k+1}). Since the integers r_k are nonnegative and strictly decreasing, the inequalities $0 \le r_{k+1} < r_k$ must eventually terminate in a remainder of 0. If we let n be the smallest integer for which $r_{n+1} = 0$, then

$$r_{n-1} = q_n r_n,$$
$$r_{n-2} = r_n - q_{n-1} q_n r_n = (1 - q_{n-1} q_n) r_n$$
$$\vdots$$

Proceeding with these calculations, one shows that r_n divides all previous r_k, and in particular a and b. Moreover, if d divides a and b then the calculations (4.1) show that d divides all successive r_k and in particular r_n. Thus r_n is the gcd of a and b.

The Python module math has a function gcd that calculates the gcd of a pair of integers. The following example shows how it works.

```
Input:
import math
g1 = math.gcd(63,27)
g2 = math.gcd(-63,27)
g3 = math.gcd(63,-27)
g4 = math.gcd(-63,-27)
print(g1,g2,g3,g4)

Output:
9 9 9 9
```

We shall not use this function but instead develop from scratch an extension of the Euclidean algorithm.

The *extended Euclidean algorithm* works like Euclidean algorithm but has two additional sequences $s_0, s_1, \ldots$ and $t_0, t_1, \ldots$. These interact with the quotient and remainder sequences as follows, where we assume that $a, b > 0$:

$$
\begin{aligned}
r_0 &= a & r_1 &= b \\
s_0 &= 1 & s_1 &= 0 \\
t_0 &= 0 & t_1 &= 1 \\
&\vdots \\
r_{k+1} &= r_{k-1} - q_k r_k, & 0 &< r_{k+1} < r_k \\
s_{k+1} &= s_{k-1} - q_k s_k, \\
t_{k+1} &= t_{k-1} - q_k t_k.
\end{aligned}
\tag{4.2}
$$

Notice that initially

$$r_0 = s_0 a + t_0 b \text{ and } r_1 = s_1 a + t_1 b.$$

Moreover, if at any stage

$$r_{k-1} = s_{k-1} a + t_{k-1} b \text{ and } r_k = s_k a + t_k b, \tag{4.3}$$

then

$$r_{k+1} = s_{k+1} a + t_{k+1} b. \tag{4.4}$$

Indeed, by (4.2)

$$r_{k+1} = r_{k-1} - q_k r_k,$$

and

$$
\begin{aligned}
s_{k+1} a + t_{k+1} b &= (s_{k-1} - q_k s_k) a + (t_{k-1} - q_k t_k) b \\
&= s_{k-1} a + t_{k-1} b - q_k (s_k a + t_k b) \\
&= r_{k-1} - q_k r_k \\
&= r_{k+1}.
\end{aligned}
$$

It follows by mathematical induction that the equation $r_k = s_k a + t_k b$ holds for all k. In particular, since the gcd of a and b is r_n, we see that the gcd may be expressed as $s_n a + t_n b$. The integers s_n and t_n are called *Bézout coefficients* of a and b.

Here is the code for the calculation of the sequences r_n, s_n, t_n. It implements the equations in (4.2).

4.3 Extended Euclidean Algorithm

```
def div_alg(a,b):
    return a // b, a%b

def extended_gcd_pos(a,b):                      # assumes a,b > 0
    r0 = a; r1 = b
    s0 = 1; t0 = 0
    s1 = 0; t1 = 1
    q = ' '; display = []                       # initialize
    while True:
        if r1 == 0:
            return [r0, s0, t0]      # gcd and Bezout coefficients
        q,r2 = div_alg(r0,r1)                   # r0 = q*r1 + r2
        s2 = s0 - q*s1
        t2 = t0 - q*t1
        r0 = r1; r1 = r2                                    # shift
        s0 = s1; s1 = s2                                    # shift
        t0 = t1; t1 = t2

-------------------------- Sample Run ---------------------------
Input:
a = 12356; b = 68
g,s,t = extended_gcd(a,b)
print(g,s,t)
-----------------------------------------------------------------
Output:
4, -7, 1272
-----------------------------------------------------------------
```

The function print_Bezout prints the Bézout equation.

```
def print_Bezout(inputlist,g,coefflist):
    Bezout = ''
    for i in range(len(inputlist)):
        coeff = tl.add_parens(str(coefflist[i]))
        integer = tl.add_parens(str(inputlist[i]))
        Bezout = Bezout + '('+ coeff +')*('+ integer +')+'
    Bezout = Bezout[0:len(Bezout)-1]            # remove last '+'
    print(str(g) + ' = ' + Bezout)

-------------------------- Sample Run ---------------------------
Input:
a = 12356
b = 68
g,s,t = extended_gcd(a,b)
inputlist = [a,b]
coefflist = [s,t]
print_Bezout(inputlist,g,coefflist
-----------------------------------------------------------------
Output:
4 = (-7)*(12356)+(1272)*(68)
-----------------------------------------------------------------
```

We can remove the restriction that a and b be positive by applying the function extended_gcd_pos to the absolute values of a and b to obtain $g = s|a| + t|b|$ and then arguing cases. For example $a < 0$ and $b > 0$ then this equation becomes $g = (-s)a + tb$, giving us the Bézout representation in this case. Here is the code:

```
def extended_gcd(a,b):      # general integers a, b; returns g,s,t
    if a == 0 and b == 0:
        return [0,0,0]                  # 0 = 0*0 +  + 0*0
    if a == 0 and b >= 0:
        return [b,0,1]                  # b = 0*0 +  + 1*b
    if a >= 0 and b == 0:
        return [a,1,0]                  # a = 1*a +  + 0*b
    if a < 0 and b < 0:
        g,s,t = extended_gcd_pos(-a,-b) # g = s(-a) + t(-b)
        return [g, -s, -t]              # g = (-s)a + (-t)b
    if a < 0 and b >= 0:
        g,s,t = extended_gcd_pos(-a,b)  # g = s(-a) + tb
        return [g, -s, t]               # g = (-s)a + tb
    if a >= 0 and b < 0:
        g,s,t = extended_gcd_pos(a,-b)  # g = sa + t(-b)
        return [g, s, -t]               # g = sa + (-t)b
    if a >= 0 and b >= 0:
        g,s,t = extended_gcd_pos(a,b)
    return [g, s, t]                    # g = sa + tb

--------------------------- Sample Run ---------------------------
Input:
a = 12356; b = -68
g,s,t = extended_gcd(a,b)
inputlist = [a,b]
coefflist = [s,t]
print_Bezout(inputlist,g,coefflist

Output:
4 = (-7)*(12356) + (-1272)*(-68)
```

4.4 Multi-extended Euclidean Algorithm

The notion of greatest common divisor may be extended to more than two numbers. For example, $\gcd(a, b, c)$ is defined as the largest common divisor of the three numbers a, b, and c. It then follows that

$$\gcd(a, b, c) = \gcd(\gcd(a, b), c)$$

4.4 Multi-extended Euclidean Algorithm

as the reader may easily check. Furthermore, there exist integers u, v, x, y such that

$$\gcd(\gcd(a,b),c) = \gcd(a,b)u + cv, \quad \gcd(a,b) = ax + by,$$

from which we obtain we obtain Bézout's equation for three integers:

$$\gcd(a,b,c) = (ax+by)u + cv = axu + byu + cv.$$

More generally, one has the formula

$$\gcd(a_1, a_2, \ldots, a_n) = \gcd(\gcd(a_1, a_2), a_3 \ldots, a_n), \qquad (4.5)$$

which may be verified by mathematical induction.

The above formulas suggests the following algorithm: First calculate $\gcd(a_1, a_2)$ and its Bezout coefficients, then calculate $\gcd(\gcd(a_1, a_2), a_3)$ and its coefficients, etc. The coefficients at each stage are calculated from those of the previous stage as in the above example. Ultimately, one winds up with the gcd and the coefficient list $[s_1, s_2, \ldots]$ related by the equation

$$\gcd(a_1, a_2, \ldots, a_n) = a_1 s_1 + a_2 s_2 + \cdots + a_n s_n. \qquad (4.6)$$

The function multi_extended_gcd(ilist) implements the algorithm recursively. The user enters the variables as a list of integers $a_1, a_2, \ldots, a_n$. The coefficients and successive gcd's are generated by the recursive function get_coeffs.

```
def multi_extended_gcd(inputlist):
    global coefflist
    n = len(inputlist)
    coefflist = ['' for x in range(n)]         # make empty list
    coefflist[0] = 1                           # initialize active part of list
    get_coeffs(inputlist)                      # implement the recursion
    g = 0
    for i in range(len(inputlist)):            # get gcd from coefficients
        g = g + coefflist[i]*inputlist[i]
    return g,coefflist

def get_coeffs(inputlist):
    global coefflist
    n = len(inputlist)
    a = inputlist[0]                           # first 2 entries of current list
    b = inputlist[1]
    G = extended_gcd(a,b)                      # latest gcd with coefficients
    x = G[1]; y = G[2]                         # coefficients of a, b
    i = 0
    while coefflist[i] != '':                  # update coefficients
        coefflist[i] = x*coefflist[i]
        i = i+1
    coefflist[i] = y                           # attach new coeff at end
    if n == 2: return coefflist                # finished
```

```
            inputlist = inputlist[1:]          # otherwise replace 1st entry
            inputlist[0] = G[0]                             # with latest gcd
            get_coeffs(inputlist)              # and do the process again

------------------------- Sample Run --------------------------
Input:
inputlist = [24,-16,20,14,-21]                     # find gcd of these
g,coefflist = multi_extended_gcd(inputlist)
print_Bezout(inputlist,g,coefflist)
---------------------------------------------------------------
Output:
1 = (-60)*(24)+(-60)*(-16)+(30)*(20)+(-10)*(14)+(-1)*(-21)
---------------------------------------------------------------
```

Here's how the scheme is carried out for the case $n = 4$:

```
---------------------------------------------------------------
# original input list              # initial coeff list
[a1,a2,a3,a4]                      [1,0,'','']

# next list                        # next list of coeffs
[a12,a3,a4]                        [x1,y1,'','']
(a12 = a1*x1+a2*y1)

[a123,a4]                          [x1*x2,y1*x2,y2,'']
(a123 = a12*x2+a3*y2= a1*x1*x2+a2*y1*x2+a3*y2)

# final list                       # final list
[a1234]                            [x1*x2*x3,y1*x2*x3,y2*x3,y3]
(a1234 = a123*x3+a4*y3 = a1*x1*x2*x3+a2*y1*x2*x3+a3*y2*x3+ a4*y3)
---------------------------------------------------------------
```

4.5 Least Common Multiple

The *least common multiple* (lcm) of a pair of positive integers m, n is the smallest positive integer k that is a multiple of both m and n. We show later that

$$\text{lcm}(m, n) \cdot \gcd(m, n) = mn. \tag{4.7}$$

For example,
$$\text{lcm}(4, 6) \cdot \gcd(4, 6) = 12 \cdot 2 = 4 \cdot 6.$$

The following function returns the lcm. It uses (4.7) with the function math.gcd.

```
def lcm(m,n):
    g = ma.gcd(m,n)
    return m*n//g
```

The *least common multiple* of a sequence of positive integers $a_1, a_2, \ldots, a_n$, denoted $\text{lcm}(a_1, a_2, \ldots, a_n)$, is the smallest positive integer that is a multiple of each a_k. The following function takes a list of integers and returns the lcm.

```
def listlcm(inputlist):
    m = inputlist[0]                          # initialize
    for k in range(1,len(inputlist)):
        m = lcm(m,inputlist[k])
    return m
------------------------- Sample Run -------------------------
Input:
print(listlcm([2,3,6,9]))

Output:
18
```

We revisit these ideas later in the context of the prime decomposition theorem, discussed in Sect. 4.7.

4.6 The Sieve of Eratosthenes

The sieve of Eratosthenes is an algorithm that generates all prime numbers up to some specified integer N. Here are the steps involved: First, make a list of all integers from 2 to N. Next, delete of all multiples of 2 larger than 2. This leaves

$$2, 3, 5, 7, 9, 11, 13, 15, 17, 19, 21, 23, 25, 27, 29, 31, 33, 35, 37, 39, 41, 43, 45, \ldots$$

Next the delete all multiples of 3 larger than 3 producing

$$2, 3, 5, 7, 11, 13, 17, 19, 23, 25, 29, 31, 35, 37, 41, 43, 47, 49, \ldots$$

Since multiples of 4 have already been deleted, the next step is the deletion all multiples of 5 except 5 itself, producing

$$2, 3, 5, 7, 11, 13, 17, 19, 23, 29, 31, 37, 41, 43, 49, \ldots$$

The process continues until only primes are left. Note that in deleting all multiples mp of a prime p, one need not consider values $m < p$, since all such multiples are either prime and so may be ignored, or they contain a prime factor $q < p$, a case that has already been dealt with. For example, to delete all multiples $2*7, 3*7, 4*7, 5*7, 6*7, \ldots$ of 7 one can start at $7*7$. The method is named after Eratosthenes of Cyrene, a Greek mathematician who lived in the third century BC.

The function sieve(N) implements the algorithm. It first creates a list marks of 1's and 0's with the property that marks(k)=1 if and only if k is prime.

```
def sieve(N):                                       # returns all primes <= N
    marks = [1 for i in range(N+1)]
    k = 2
    while (k*k <= N):
        if marks[k] == 1:                           # if k is prime
            for i in range(k*k,N+1,k):  # mark by 0 all the multiples
                marks[i] = 0                        # k*k, (k+1)*k,...
        k += 1
    primes = []
    for k in range(2,len(marks)):
        if marks[k] == 1:
            primes.append(k)                        # attach prime
    return primes
```

---------------------------- Sample Run ----------------------------
Input:
print(sieve(500))

Output:
[2, 3, 5, 7, 11, 13, 17, 19, 23, 29, 31, 37, 41, 43, 47, 53, 59, 61,
67, 71, 73, 79, 83, 89, 97, 101, 103, 107, 109, 113, 127, 131, 137,
139, 149, 151, 157, 163, 167, 173, 179, 181, 191, 193, 197, 199,
211, 223, 227, 229, 233, 239, 241, 251, 257, 263, 269, 271, 277,
281, 283, 293, 307, 311, 313, 317, 331, 337, 347, 349, 353, 359,
367, 373, 379, 383, 389, 397, 401, 409, 419, 421, 431, 433, 439,
443, 449, 457, 461, 463, 467, 479, 487, 491, 499]

4.7 The Fundamental Theorem of Arithmetic

The theorem in the heading, also called the *unique prime factorization theorem*, asserts that every integer $N \geq 2$ may be written uniquely as a product of the form

$$N = p_1^{e_1} p_2^{e_2} \cdots p_m^{e_m},$$

4.7 The Fundamental Theorem of Arithmetic

where the p_j's are prime, $p_1 < p_2 \ldots < p_m$, and the e_j's are positive integers. The theorem may be established by mathematical induction. The following function returns the prime factorization of N.

```
def prime_factorization(N):
    prime_list = sieve(N//2+1)       # get primes less than N/2 + 1
    exponents = []
    primes = []
    for i in range(len(prime_list)): # run through the primes
        p = prime_list[i]
        e = 0                         # initialize exponent for this prime
        while N % p == 0:             # if prime p divides N,
            N = N/p                   # keep dividing it out,
            e = e+1                   # and update the exponent.
        if e != 0:                    # if p divided N,
            primes.append(p)          # include it,
            exponents.append(e)       # and its exponent.
    return primes, exponents

def print_factorization(primes,exponents):
    P = ''                            # string for prime factorization
    for i in range(len(primes)):
        p = primes[i]
        e = exponents[i]
        if e > 1:
            P = P + '('+ str(p) +'^'+ str(e) + ')*'
        else:
            P = P + str(p) + '*'
    print(P[:len(P)-1])

------------------------- Sample Run ---------------------------
Input:
N = 300042
primes,exponents = prime_factorization(N)
print(primes,exponents)
print_factorization(primes,exponents)
----------------------------------------------------------------
Output:
[2, 3, 79, 211] [1, 2, 1, 1]
2 * 3^2 * 79 * 211
----------------------------------------------------------------
```

The Fundamental Theorem of arithmetic has an interesting connection with the notions of greatest common divisor and least common multiple of positive integers a, b. To see this, write their prime decompositions as

$$a = q_1^{d_1} q_2^{d_2} \cdots q_n^{d_n} \text{ and } b = q_1^{e_1} q_2^{e_2} \cdots q_n^{e_n},$$

where q's are all the primes in a and b, but where now some of the exponents may be zero. It is not hard to verify that

$$\gcd(a,b) = q_1^{x_1} q_2^{x_2} \cdots q_n^{x_n} \text{ and } \mathrm{lcm}(a,b) = q_1^{y_1} q_2^{y_2} \cdots q_n^{y_n}$$

where $x_j = \min(d_j, e_j)$ and $y_j = \max(d_j, e_j)$. Since $x_j + y_j = d_j + e_j$ (argue by cases), it follows that $\gcd(a,b) \cdot \mathrm{lcm}(a,b) = ab$, as asserted earlier.

4.8 Modular Arithmetic

Let m, a, and b be integers with $m \geq 2$. Then a is said to be *congruent to b modulo m*, written

$$a \equiv b \pmod{m} \text{ or } a \equiv_m b$$

if m divides $a - b$. For example,

$$17 \equiv 3 \pmod{7}, \quad -17 \equiv 1 \pmod{6}, \text{ and } -17 \equiv -2 \pmod{5}.$$

Note that if r is the remainder upon dividing a by m, that is,

$$a = qm + r, \quad 0 \leq r < m,$$

then $a - r = qm$. Thus every integer a is congruent to one of the numbers $0, 1, \ldots, m-1$ modulo m.

For the rest of this section we fix $m \geq 2$ and denote by $R(a)$ the remainder on division of m by a. The function R has the following properties:

(a) $R(R(a)) = R(a)$.
(b) $R(ms) = 0$.
(c) $R(a + ms) = R(a)$.
(d) $R(a + b) = R(R(a) + R(b))$.
(e) $R(ab) = R(R(a)R(b))$.

Part (a) simply asserts that $R(r) = r$ for any r with $0 \leq r < m$, and (b) says that multiples of m have zero remainders. For the remaining properties set

$$a = pm + R(a), \ (0 \leq R(a) < m) \text{ and } b = qm + R(b), \ (0 \leq R(b) < m).$$

Part (c) then follows from $a + ms = (p + s)m + R(a)$, which shows that $a - R(a)$ is a multiple of m.

For part (d) we have

$$a + b = (p + q)m + R(a) + R(b).$$

4.8 Modular Arithmetic

But also,
$$a + b = sm + R(a+b)$$

Thus the right sides are equal:
$$(p+q)m + R(a) + R(b) = sm + R(a+b)$$

Therefore,
$$R\big((p+q)m + R(a) + R(b)\big) = R\big(sm + R(a+b)\big).$$

This equality reduces to (d) by part (c).

Finally for part (e) we have
$$ab = pqm^2 + R(a)qm + R(b)pm + R(a)R(b) = tm + R(ab),$$

from which it follows that $R(a)R(b) = R(ab) + um$ for some integer u. Taking R of both sided and using (c) completes the proof.

The set of remainders $0, 1, \ldots, m-1$ is denoted by $\mathbb{Z}_m$. We define addition $+_m$, and multiplication $*_m$ on $\mathbb{Z}_m$ by
$$a +_m b = R(a+b), \quad a *_m b = R(ab).$$

The operations have the following properties:

- associative laws:
$$(a +_m b) +_m c = a +_m (b +_m c), \text{ and } (a *_m b) *_m c = a *_m (b *_m c).$$

- commutative laws:
$$a +_m b = b +_m a, \text{ and } a *_m b = b *_m a.$$

- existence of identities:
$$a +_m 0 = 0, \quad a *_m 1 = a.$$

- distributive law:
$$(a +_m b) *_m c = a *_m b +_m a *_m c.$$

- additive inverse:
  For each $a \in \mathbb{Z}_m$ there exists a unique $b \in \mathbb{Z}_m$ such that $a +_m b = 0$.
- multiplicative inverse:
  If m and $a \in \mathbb{Z}_m$ are relatively prime then exists a unique $b \neq 0 \in \mathbb{Z}_m$ such that $a *_m b = 1$. In particular, If m is prime, then for each $a \neq 0 \in \mathbb{Z}_m$ there exists a unique $b \neq 0 \in \mathbb{Z}_m$ such that $a *_m b = 1$.

For example, for the distributive law we have

$$\begin{aligned}
(a +_m b) *_m c &= R((a +_m b)c) & &\text{(definition of } *_m) \\
&= R(R(a+b)c) & &\text{(definition of } +_m) \\
&= R(R(a+b)R(c)) & &\text{(property (a) of } R) \\
&= R((a+b)c) & &\text{(property (e) of } R) \\
&= R(ac+bc) & &\text{(integer distributive law)} \\
&= R(R(ac)+R(bc)) & &\text{(property (d) of } R) \\
&= R((a *_m c)+(b *_m c)) & &\text{(definition of } *_m) \\
&= (a *_m c) +_m (b *_m c) & &\text{(definition of } +_m)
\end{aligned}$$

For the multiplicative inverse property note that if m and $a \in \mathbb{Z}_m$ are relatively prime then $\gcd(a, m) = 1 = xa + ym$, where x, y are the Bézout coefficients of a and m. Thus

$$1 = R(1) = R(xa + ym) = R(xa) = R(R(x)R(a)).$$

Setting $b = R(x)$ and noting that $R(a) = a$, we have

$$1 = R(ba) = R(ab) = a *_m b.$$

Here are functions that generate addition and multiplication tables for a user-entered modulus.

```
def addition_modtable(m):
    table = []
    header = []
    for k in range(m):
        header = header+[str(k)]
    header = ['+'] + header                       # top labels
    table = [header]
    for i in range(m):                            # body of table
        row = []
        for j in range(m):
            row = row + [str((i+j)%m)]
        row = [str(i)] + row                      # left labels
        table = table + [row]
    return table

def multiplication_modtable(m):
    table = []
    header = []
    for k in range(m):
        header = header+[str(k)]
    header = ['*'] + header
    table = [header]
    for i in range(m):
        row = []
```

4.8 Modular Arithmetic

```
            for j in range(m):
                row = row + [str((i*j)%m)]
            row = [str(i)] + row
            table = table + [row]
        return table
```

```
--------------------------- Sample Run ---------------------------
Input:
table = addition_modtable(5)              # mod 5 addition table
print('mod 5')
tl.format_print(table, 3, 'left')
print('\n')
table =multiplication_modtable(5)         # mod 5 addition table
tl.format_print(table, 3, 'left')
print('\n')
table = addition_modtable(4)              # mod 4 addition table
print('mod 4')
tl.format_print(table, 3, 'left')
print('\n')
table = multiplication_modtable(4)        # mod 4 addition table
tl.format_print(table, 3, 'left')
------------------------------------------------------------------
Output:
mod 5
+   0   1   2   3   4
0   0   1   2   3   4
1   1   2   3   4   0
2   2   3   4   0   1
3   3   4   0   1   2
4   4   0   1   2   3

*   0   1   2   3   4
0   0   0   0   0   0
1   0   1   2   3   4
2   0   2   4   1   3
3   0   3   1   4   2
4   0   4   3   2   1

mod 4
+   0   1   2   3
0   0   1   2   3
1   1   2   3   0
2   2   3   0   1
3   3   0   1   2

*   0   1   2   3
0   0   0   0   0
1   0   1   2   3
2   0   2   0   2
3   0   3   2   1
------------------------------------------------------------------
```

Here are functions that return the additive and multiplicative inverses of an integer a.

```
def mod_add_inv(a,m):
    a = a%m
    for b in range(0,m):
        if (a+b)%m  == 0:
            break
    return b

def mod_mult_inv(a,m):
    for b in range(1,m):
        x = a*b
        if (x%m)  == 1:
            return b
    return -1         # no inverse
```

--------------------------- Sample Run ---------------------------
Input:
print(mod_add_inv(40,7))
print(mod_mult_inv(678,7))
print(mod_mult_inv(678,6))

Output:
2
6
no inverse

Arithmetic   5

In this chapter we construct the module `Arithmetic.py`, which evaluates arithmetic expressions with *Gaussian rational numbers*, that is, complex numbers $z = a + bi$, where a and b are rational numbers, called the *real* and *imaginary parts* of z, respectively. (Arithmetic expressions were defined in Chap. 2.) Complex numbers are stored in the module as lists. For example, the complex number $2/3 + (4/5)i$ is stored as `[2/3,4/5]`. All quantities are in string form. The module is headed by the following import statement:

```
--------------------------- Arithmetic.py ---------------------
import Tools as tl
import math as ma
---------------------------------------------------------------
```

5.1 The Main Function

The function `main` takes an arithmetic expression and returns a formatted complex number. Here's a sample run:

```
----------------------------------------------------------------
Input:
expr =  '(3.2/5i) + 4(7.1i + 2.5/3)^3 - 1.7'
print(main(expr)[0])
----------------------------------------------------------------
Output:
-135941/270-(514919/375)i
----------------------------------------------------------------
```

A companion function, `decimal_approx(fraction,p)`, returns a decimal approximation of the fraction that is accurate to *p* decimal places. The number of decimal places for the approximation can be quite large. For example, applying `main` to the expression (2/7)^10*(1/5^8) yields the fraction 1024/110341894140625. Applying `decimal_approx` to this fraction with p = 40 yields the decimal approximation

$$.0000000000092802467093320448759034459688$$
$$= 9.280246709332044875903445968 * 10^{\wedge}(-12)$$

The second form is in so-called *scientific notation*. By contrast, the Python function `eval` applied to the above expression (with the caret symbol ^ replaced by the double asterisk **) returns only a decimal, in this case 9.280246709332041e-12, considerably less accurate.

Here is the code for `main`. The supporting functions are developed in the remaining sections.

```
def main(expr):
    # input: complex arithmetic expression
    # output: formatted complex number z=a+bi and list c=[real,imag]
    global idx              # points to characters in the string expr
    expr = expr.replace(' ','')            # remove extra spaces
    expr = tl.fix_signs(expr)
    expr = tl.fix_operands(expr)
    expr = tl.insert_asterisks(expr,'')    # no variables
    idx = 0                                # start
    c = allocate_ops(expr,0)   # does the calculations, returns list
    z = pair2complex(c)    # convert list c into formatted complex no.
    return z,c
```

5.2 Conversion Functions

The following function takes a decimal (a string comprised of symbols from .0123456789) and converts it to a fraction.

```
def decimal2frac(numeric):
    if '.' not in numeric: return numeric          # not a decimal
    whole, decimal = numeric.split('.')    # e.g. 12.34 --> 12,34
    zeros = '0'*(int(len(decimal)))                # '00'
    denominator = '1' + zeros                      # 100
    numerator = whole + decimal                    '1234'
    return numerator + '/' + denominator           # 1234/100
```

5.2 Conversion Functions

The function frac2intpair(r) takes a string fraction r and returns the numerator and denominator as integers.

```
def frac2intpair(r):                              # r = 'm/n'
    if '/' not in r: r = r + '/1'  # make r a fraction for uniformity
    m,n = r.split('/')
    return int(m),int(n)           # integer numerator and denominator
```

The function pair2frac(num,den) does the reverse of frac2intpair(r). It takes a pair of integers num,den and returns the fraction 'num/den', suitably formatted and reduced to lowest terms.

```
def pair2frac(num,den):
    # returns fraction num/den reduced
    if  num == 0: return '0'
    if  den < 0: num = -num; den = -den      # put minus sign on top
    if  num == den: return '1'
    if  num == -den: return '-1'
    if  den == 1: return str(num)            # no denominator needed
    num,den = reduce_int_pair(num,den)       # cancel common factor
    return str(num) + '/' + str(den)              # combine parts

def reduce_int_pair(a,b):
    c = ma.gcd(a,b)             # largest common factor of a and b
    return a//c, b//c                           # integer division
------------------------------ Sample Run --------------------------
Input:
print(pair2frac(40,-555))

Output:
-8/111
```

The function numeric2list(r) takes a purely real or purely imaginary numeric *r* and returns the double list that represents it.

```
def numeric2list(r):
    if r == 'i':  return ['0','1']
    if r == '-i': return ['0','-1']
    if 'i' not in r:
        return [decimal2frac(r),'0']              # purely real
    r = r.replace('i','')
    return ['0', decimal2frac(r)]                 # purely complex
------------------------------ Sample Run --------------------------
```

```
Input:
print(numeric2list('2.3'))
print(numeric2list('.56i'))
```

```
Output:
['23/10', '0']
['0', '56/100']
```

The function pair2complex(c) takes a pair c = [a,b] of fractions a,b representing a complex number and returns a formatted complex number a+bi.

```
def pair2complex(c):              # c = [a, b], a, b string fractions
    real = c[0]; imag = c[1]
    if '/' in real:
        num,den = real.split('/')
        real = pair2frac(int(num),int(den))     # format real part
    if '/' in imag:
        num,den = imag.split('/')
        imag = pair2frac(int(num),int(den))  # format imaginary part
        imag = '(' + imag + ')'
    if real != '0':
        imag = imag.replace('(-','-(')         # pull '-' outside paren

    #### special cases:
    if real == '0' and imag == '0': return '0'
    if real == '0' and imag == '1': return 'i'
    if real == '0' and imag == '-1': return '-i'
    if imag == '0': return real
    if real == '0': return imag + 'i'
    if imag == '1':  imag = ''            # coefficient 1 not needed
    if imag == '-1': imag = '-'

    #### general case:
    c = real + '+' + imag + 'i'
    c = tl.fix_signs(c)
    return c
```
--------------------------- Sample Run ---------------------------
```
Input:
c = ['12/34','-1/5']
print(formatcomplex(c))
```

```
Output:
6/17-(1/5)i
```

5.3 Arithmetic Operations on Fractions

The functions in this section perform arithmetic calculations on fractions represented as strings r='m/n', s='p/q', where m, n, p, q are integers.

```
def frac_sum(r,s):                          # r + s = m/n + p/q = (mq+np)/nq
    m,n = frac2intpair(r)                   # convert fraction to integer pair
    p,q = frac2intpair(s)
    return pair2frac(m*q + n*p, n*q)        # return reduced fraction

def frac_diff(r,s):
    t = '-'+s                               # make s negative
    t = tl.fix_signs(t)                     # remove extraneous '+', '-'
    return frac_sum(r,t)

def frac_prod(r,s):                         # r*s = m/n * p/q = m*p/n*q
    m,n = frac2intpair(r)
    p,q = frac2intpair(s)
    return pair2frac(m*p,n*q)

def frac_recip(s):
    p,q = frac2intpair(s)
    return pair2frac(q,p)                                      # q/p

def frac_quo(r,s):
    t = frac_recip(s)                                          # t = q/p
    return frac_prod(r,t)                                      # (m/n)*(q/p)

def frac_power(s,exp):                                         # (p/q)^exp
    if exp == 0: return '1'
    t = s;                                                     # default
    if exp < 0:
        exp = -exp                          # make exponent positive
        t = frac_recip(t)                                      # invert t
    num,den = frac2intpair(t)
    num,den = num**exp, den**exp            # raise both to positive power
    return pair2frac(num,den)                                  # reduced m/n

---------------------------- Sample Run ----------------------------
Input:
r = '2/3'; s = '-5/7'; exp = -4
print('r+s   = ', frac_sum(r,s))
print('r-s   = ', frac_diff(r,s))
print('r*s   = ', frac_prod(r,s))
print('r/s   = ', frac_quo(r,s))
print('r^exp = ', frac_power(r,exp))
--------------------------------------------------------------------
```

```
Output:
r+s   = -1/21
r-s   = 29/21
r*s   = -10/21
r/s   = -14/15
r^exp = 81/16
```

5.4 Complex Operations

Arithmetic operations on complex numbers are defined by

$$(a+bi) \pm (c+di) = (a+c) \pm (b+d)i,$$
$$(a+bi)(c+di) = (ac-bd) + (bc+ad)i,$$
$$\frac{a+bi}{c+di} = \frac{(a+bi)(c-di)}{c^2+d^2} = \frac{ac+bd}{c^2+d^2} + \frac{bc-ad}{c^2+d^2}i.$$

The following functions implement these operations. They operate on pairs u = [a,b] and v = [c,d], representing complex numbers $a+bi$ and $c+di$, respectively, where a, b, c, d are fractions, and return a list [real,imag] of real and imaginary parts of the computation.

```
def complex_sum(u,v):           # u = [a,b], v = [c,d], a,b,c,d fractions
    a,b = u[0],u[1]             # a = m/n, b = p/q, m,n,p,q integers
    c,d = v[0],v[1]
    real = frac_sum(a,c)        # add real parts: a+c
    imag = frac_sum(b,d)        # add imag parts: b+d
    return [real,imag]          # return string list

def complex_diff(u,v):
    a,b = u[0],u[1]
    c,d = v[0],v[1]
    real = frac_diff(a,c)
    imag = frac_diff(b,d)
    return [real,imag]

def complex_prod(u,v):
    a,b = u[0],u[1]
    c,d = v[0],v[1]
    ac = frac_prod(a,c)
    bd = frac_prod(b,d)
    ad = frac_prod(a,d)
    bc = frac_prod(b,c)
```

5.4 Complex Operations

```
        real = frac_diff(ac,bd)                    # ac - bd, ad + bc
        imag = frac_sum(ad,bc)
        return [real,imag]

    def complex_recip(u):
        a,b = u[0],u[1]
        if a == 0: return ['0','-'+frac_recip(b)]  # ['0',-1/b]
        if b == 0: return [frac_recip(a),'0']      # [1/a,'0']
        a2 = frac_power(a,2)
        b2 = frac_power(b,2)
        a2_plus_b2 = frac_sum(a2,b2)
        real = frac_quo(a,a2_plus_b2)              # a/(a^2 + b^2)
        imag = frac_quo(b,a2_plus_b2)
        imag = frac_diff('0',imag)                 # -b/(a^2 + b^2)
        return [real,imag]           # a/(a^2 + b^2) - i a/(a^2 + b^2)

    def complex_quo(u,v):
        w = complex_recip(v)
        return complex_prod(u,w)

    def complex_power(u,exp):                                   # u^exp
        if exp == 0: return ['1','0']
        if  exp < 0:
           exp = -exp
           u = complex_recip(u)
        v = u
        for k in range(exp-1):    # multiply u times itself exp-1 times
            v = complex_prod(u,v)
        return v

    ---------------------------- Sample Run ----------------------------
    Input:
    u = ['2/3','4/5']
    v = ['2','-7']
    print('sum   = ',complex_sum(u,v))
    print('diff  = ',complex_diff(u,v))
    print('prod  = ',complex_prod(u,v))
    print('recip = ',complex_recip(v))
    print('quo   = ',complex_quo(u,v))
    print('power = ',complex_power(u,3))
    print('power = ',complex_power(u,-3))
    --------------------------------------------------------------------
    Output:
    sum   =  ['8/3', '-31/5']
    diff  =  ['-4/3', '39/5']
    prod  =  ['104/15', '-46/15']
    recip =  ['2/53', '7/53']
    quo   =  ['-64/795', '94/795']
    power =  ['-664/675', '208/375']
    power =  ['-1400625/1815848', '-394875/907924']
    --------------------------------------------------------------------
```

5.5 The Allocator

The function allocate_ops(expr,mode) calls on the preceding functions for the computational tasks. It takes an arithmetic expression expr and returns a complex number in the form pair of simple fractions. The function operates recursively, calling itself to perform the operation. The parameter mode is used to enforce hierarchy of operations: addition and subtraction last. The following examples illustrate the hierarchy.

$$3/4*2 = 3/8, \quad 3*4/2 = 6, \quad 2/3\hat{\ }4 = 2/81,$$
$$2\hat{\ }3/4 = 2, \quad 2*3-4 = 2, \quad 2-3*4 = -10$$

```
def allocate_ops(expr,mode):           # returns a pair of fractions
    global idx
    while idx < len(expr):
        ch = expr[idx]                 # character at index idx
        if ch in '.0123456789i':       # beginning of a numeric
            start = idx
            r,idx = tl.extract_numeric(expr, start)
            r = numeric2list(r)
        elif ch == '+':
            if mode > 0: break         # wait for higher mode
            idx += 1
            s = allocate_ops(expr,0)
            r = complex_sum(r,s)
        elif ch == '-':
            if mode > 0: break         # wait for higher mode
            idx += 1
            s = allocate_ops(expr,1)
            r = complex_diff(r,s)
        elif ch == '*':
            if mode > 1: break
            idx += 1
            s = allocate_ops(expr,1)
            r = complex_prod(r,s)
        elif ch == '^':
            idx += 1
            exp = allocate_ops(expr,2)[0]        # get exponent
            r = complex_power(r, int(exp))
        elif ch == '/':
            if mode > 1: break
            idx += 1
            s = allocate_ops(expr,1)             # get denominator
            r = complex_quo(r,s)
        elif ch == '(':
            idx = idx + 1                        # skip '('
            r = allocate_ops(expr,0)    # calculate stuff inside ()
            idx = idx + 1                        # skip ')'
        elif ch == ')': break
    return r
```

Fig. 5.1 Recursion diagram for $a+b*c\hat{~}d$

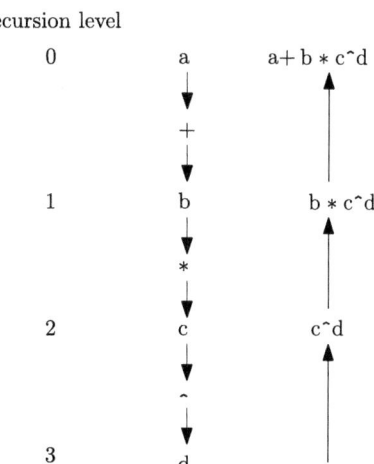

To see how the allocator works, consider the expression 'a+b*c+d^e'. The function, initially in mode 0, reads a, then goes to '+' and calls itself to retrieve b, goes on to '*' and calls itself to retrieve c, goes on to '^' and calls itself to retrieve d, comes back to calculate c^d, etc. The above diagram traces the journey (Fig. 5.1).

5.6 Rendering a Fraction into a Decimal

The function frac_decimal_approx(fraction,p) below takes a fraction and a positive integer p and calculates a decimal approximation of the fraction to p places. To see how the function works, suppose we wish to approximate the fraction 45/67 with a decimal that is accurate to $p=4$ places. To achieve this we first multiply the numerator by 10^4 and then apply the division algorithm to 450000/67 to obtain

$$450000 = 6716 * 67 + 43.$$

Dividing both sides by 10000*67 yields

$$\frac{45}{67} = 0.6716 + \frac{43}{67*10000} = 0.6716 + \frac{43}{67}*10^{-4}.$$

The number on the extreme right is less than $10^{-4} = 0.0001$, hence .6716 agrees with 45/67 in 4 decimal places and so is the desired approximation.

In general, to approximate the integer fraction a/b to p decimal places, apply the division algorithm as follows:

$$a*10^p = q*b+r, \quad 0 \le r < b.$$

Divide both sides by $b * 10^p$:

$$\frac{a}{b} = \frac{q}{10^p} + \left(\frac{r}{b}\right) * 10^{-p}$$

Since $r/b < 1$, we have $(r/b) * 10^{-p} < 10^{-p}$, and so a/b and $q/10^p$ agree in p decimal places.

The result 0.6716 in the above example is achieved in the program by taking the quotient $q = 6717$ and suitably positioning the decimal point, in this case four places to the left. In general, the positioning depends on both of the integer strings p and q. Here is the code:

```
def frac_decimal_approx(fraction,p):          # fraction = a/b
    # input: ratio of integers
    # output: approximatoin to p places
    if '/' not in fraction:
        return fraction
    a,b = frac2intpair(fraction)
    if a == b:
        return '1'                              # trivial case
    c = a*10**p
    q,r = c//b, c%b                             # div alg for c,b
    q = str(q); r = str(r)
    L = len(q)

        # place the decimal point p places to left in q:
    if p < L:
        approx = q[:L-p]+ '.' + q[L-p:]
    if p == L:
        approx =   '.' + q
    if p > L:
        zerostring = (p-L)*'0'
        approx = '.' + zerostring + q
    L = len(approx)
    for i in range(L-1,-1,-1):                  # remove trailing zeros
        if approx[i] != '0':
            break            # i now at first nonzero before zero trail
    approx = approx[:i+1]                       # chop off zero trail
    if approx[len(approx)-1] == '.':            # if last symbol is '.'
        approx = approx.replace('.','')         # remove it
    return approx
-------------------------- Sample Run --------------------------
Input:
fraction = '45/67'
print(frac_decimal_approx(fraction,4))
print(frac_decimal_approx(fraction,50))

Output:
.6716
.67164179104477611940298507462686567164179104477611
```

The function decimal_approx(z,p) generalizes frac_approx by allowing arbitrary Gaussian rationals as input. It simply applies frac_approx to the real and imaginary parts of z and returns a formatted complex number with the approximation of each part.

```
def decimal_approx(z,p):
    # input: z = complex number, p = number of decmal places
    # output: decimal complex number z and corresponding list d
    c = main(z)[1]                       # get real and imaginary parts of z
    real = frac_decimal_approx(c[0],p)   # approximate real
    imag = frac_decimal_approx(c[1],p)   # and imaginary parts
    d = [real,imag]
    w = pair2complex(d)                  # format number
    return w,d                           # number, list form
```
------------------------- Sample Run ---------------------------
Input:
c = '76/7-(151/29)i'
print(decimal_approx(c,21)[0])

Output:
10.857142857142857142857-5.2068965517241379310345i

5.7 Converting to Scientific Notation

The following function takes a decimal and renders it into scientific notation: a decimal consisting of a single digit whole part, a fractional part, and a suitable power of 10 to compensate for moving the decimal point. The sample run illustrates this.

```
def scientific_notation(decimal):
    if '.' not in decimal:
        D = len(decimal)
        if D == 1:                                     # 3 --> 3
            return decimal
        if D == 2:                                     # 33 --> 3.3 * 10
            return tl.insert_string(decimal,'.',1)[0] + '*' + '10'
        if D > 2:                                      # 333 --> 3.33 * 10^2
            return tl.insert_string(decimal,'.',1)[0] \
                + '*' + '10^' + str(D - 1)
    left,right = decimal.split('.')
    if left != '' and main(left)[0] == '0':
        left = ''
    if left == '':
        R = len(right)
        right = right.replace('.','')
```

```
            if R == 1:                       # .3 --> 3*10^(-1)
                return right + ' *10^(-1)'
            if R > 1:                        # .33 --> 3.3*10^(-1)
                return tl.insert_string(right,'.',1)[0] \
                             + '*' + '10^(-1)'
    L = len(left)
    if L == 1:                               # 3.3 --> 3.3
        return decimal
    decimal = decimal.replace('.','')
    if L == 2:                               # 33.3 --> 3.3*10
        return tl.insert_string(decimal,'.',1)[0] + '*' + '10'
    exp = str(L-1)
            # 333.3 --> 3.3*10^2:
    return tl.insert_string(decimal,'.',1)[0] +'*'+'10^'+ str(exp)

--------------------------- Sample Run ---------------------------
Input:
dlist = ['1','12','123','.1','.12','.123','1.2', '12.3','123.4']
ddlist = []
for d in dlist:    # make a list of pairs [d,sci] for format print
    sci = scientific_notation(d)
    ddlist.append([d,'=',sci])
tl.format_print(ddlist, 2, 'left')
------------------------------------------------------------------
Output:
12      =   1.2*10
123     =   1.23*10^2
.1      =   1 *10^(-1)
.12     =   1.2*10^(-1)
.123    =   1.23*10^(-1)
1.2     =   1.2
12.3    =   1.23*10
123.4   =   1.234*10^2
------------------------------------------------------------------
```

5.8 Evaluating an Expression

The following function takes a numerical value and an algebraic expression containing a variable and evaluates the expression at that value, returning both a Gaussian rational and an approximation.

```
------------------------------------------------------------------
def evaluate(expr,varval,p):
    var = tl.get_var(expr)
    if var == '':
        return main(expr)[0],''
    e = expr.replace(var,'(' + varval + ')')
    e = tl.fix_signs(e)
```

5.8 Evaluating an Expression

```
    e = main(e)[0]
    if p == '': return e,''
    return e, decimal_approx(e,p)[0]
```

```
--------------------------- Sample Run ---------------------------
Input:
expr = '(2x^2 + x + 4)^2/(1+(1/2i)x^2 - (7/3)x - 11/8)/(x^2+1)'
var_val = '1.1'
p = 9
expr_val, dec_approx = evaluate(expr,var_val,p)
print('expression value: ', expr_val)
print('decimal approx:   ', dec_approx)
------------------------------------------------------------------
Output:
expression value:   -8271874416/202937125+(8506205136/1014685625)i
decimal approx.:    -40.76077463+8.383094158i
------------------------------------------------------------------
```

Here is the analog for expressions with integers modulo a positive integer m.

```
------------------------------------------------------------------
def mod_evaluate(expr,varval,m):
    var = tl.get_var(expr)
    exprval = expr.replace(var,'(' + varval + ')')
    exprval = tl.fix_signs(exprval)
    return int(main(exprval)[0])%
------------------------------------------------------------------
Input:
expr = '(3x-57)^177'
var_val = '2'
for m in range(2,14):
    expr_val = mod_evaluate(expr,var_val,m)
    print('mod',m,' ',expr_val)

--------------------------- Sample Run ---------------------------
Output:
mod 2   1
mod 3   0
mod 4   1
mod 5   4
mod 6   3
mod 7   6
mod 8   5
mod 9   0
mod 10  9
mod 11  5
mod 12  9
mod 13  1
------------------------------------------------------------------
```

5.9 Ordering Fractions

Because fractions are written as strings, one needs a special function to compare them. Since main returns a negative fraction with the negative sign attached to the numerator, as in -2/3, the denominator of a fraction is always positive. It follows that for any pair fractions a_1/b_1 and a_2/b_2 returned by main,

$$\frac{a_1}{b_1} < \frac{a_2}{b_2} \text{ if and only if } a_1 b_2 < a_2 b_1.$$

The function min_max_frac uses this fact in determining which of two fractions is the larger, returning them in order of size, the smaller first.

```
def min_max_frac(frac1,frac2):
    fr1 = frac1; fr2 = frac2
    if '/' not in fr1:
        fr1 = fr1 + '/1'                     # for uniformity
    if '/' not in fr2:
        fr2 = fr2 + '/1'
    a1,b1 = fr1.split('/')                   # a1/b1, b1>0
    a2,b2 = fr2.split('/')                   # a2/b2, b2>0
    a1 = int(a1); b1 = int(b1)    # convert strings into integers
    a2 = int(a2); b2 = int(b2)
    if  a1*b2 < a2*b1:
        return frac1,frac2
    return frac2,frac1
```

The function is_less(frac1,frac2) returns True if frac1<=frac2 and False otherwise. It will be used below in ordering a list of fractions. The function absval returns the absolute value of a numeric.

```
def is_less(frac1,frac2):
    minfrac,maxfrac = min_max_frac(frac1,frac2)
    return minfrac == frac1

def absval(a):
    b = main(a)[0]          # convert possible decimal into fraction
    if is_less(b,'0'):
        return tl.fix_signs'-'+ a
    else:
        return a
```

The function `frac_sort(frac_list)` takes a list of fractions and returns the list sorted from smallest to largest. It uses the so-called *bubble sort* algorithm, where the "lighter" values "float" to the top. The remaining functions use `frac_sort` to find the indices of the smallest and largest value in a list.

```
def frac_sort(frac_list):                              # min to max
    L = len(frac_list)
    for i in range(L):
        for j in range(i+1,L):
            if is_less(frac_list[j],frac_list[i]):
                swap = frac_list[i]                    # put f(i) before f(j)
                frac_list[i] = frac_list[j]
                frac_list[j] = swap
    return frac_list

def index_of_min(frac_list):
    sorted_list = frac_sort(frac_list)                 # small to large
    return frac_list.index(sorted_list[0])             # index of smallest no.

def index_of_max(frac_list):
    L = len(frac_list)
    sorted_list = frac_sort(frac_list)
    return frac_list.index(sorted_list[L-1])           # index of largest no.

--------------------------- Sample Run ---------------------------
Input:
frac_string = '43/46,-81/83,-64/67,-33/34,671/678,52/55,111/123,7/77'
frac_list = frac_string.split(',')
print(frac_sort(frac_list))
idx_min = index_of_min(frac_list)
idx_max = index_of_max(frac_list)
print(frac_list[idx_min], frac_list[idx_max])

Output:
['-81/83','-33/34','-64/67','7/77','11/12','43/46', 52/55','671/678']
-81/83   671/678
```

5.10 Application: Roots by Interval Halving

A *zero* or *root* of a function f is a value z such that $f(z) = 0$. The *interval halving method* for finding approximate roots applies to continuous functions f, that is, functions that have no gaps or jumps in their graphs. The method finds an approximate value of a zero of f located in a given interval $[a, b]$. The algorithm begins by determining the sign of the product $f(a)f(b)$. If the sign is negative then the graph of f crosses the x axis somewhere between a and b, indicating that the interval contains a zero of f. Dividing the interval into two pieces

Fig. 5.2 Interval Halving method

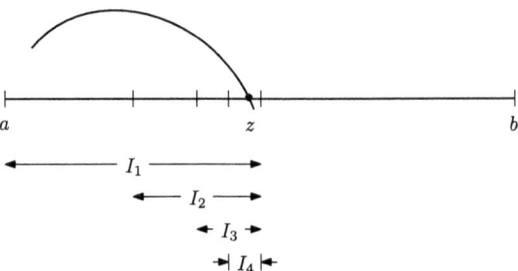

$[a, c]$ and $[c, b]$ and checking the signs of $f(a)f(c)$ and $f(c)f(b)$ indicates which of these intervals has a zero. The process continues up to any desired degree of accuracy, measured by the lengths of the bisected intervals. If an interval length is less than a predetermined value, then the desired zero may be approximated by any member in the interval. Figure 5.2 illustrates the idea. The interval halving method may miss some zeros. To approximate these one may need to adjust the initial interval $[a, b]$.

The function interval_halving(f,a,b,accuracy) implements the procedure for algebraic expressions $f(x)$.

```
def interval_halving(f, a, b, accuracy):      # f,a,b as strings
    accuracy = main(accuracy)[0]              # convert to fraction form
    while True:
        # midpoint of current interval:
        c = main('('+ a + '+' + b + ')/2')[0]
        cval = evaluate(f,c,'')               # expression values
        aval = evaluate(f,a,'')
        bval = evaluate(f,b,'')
        prod = main('('+ aval +')('+ bval + ')')[0]
        if is_less('0',prod): return 'none'   # no zero in [a,b]
        if aval == '0': return a              # special cases
        if bval == '0': return b
        if cval == '0': return c
        prod = main('('+ aval +')('+ cval + ')')[0]
        if is_less(prod,'0'):                 # a zero between a and c
            b = c                             # new interval is left half
        prod = main('('+ cval +')('+ bval + ')')[0]
        if is_less(prod,'0'):                 # a zero between c and b
            a = c                             # new interval is right half
        z = c                                 # approximate zero
        interval_length = main(b + '-('+ a +')')[0]
        if is_less(interval_length,accuracy):
            break
    return z
```

Input:
f = '(2.4-x^2)(x^2+x+1)^2'; a = '.5'; b = '2'; p = 50
for i in range(10):

5.10 Application: Roots by Interval Halving

```
        zeros = '0'*i
        accuracy = '.'+ zeros + '1'
        p = 50
        approx_root = interval_halving(f, a, b, accuracy)
        if approx_root == 'none':
            print('no root')
            break
        dec_approx = frac_approx(approx_root,p)
        print('accuracy:              ', accuracy)
        print('approximate root:      ', approx_root)
        print('decimal approximation:', dec_approx,'\n')
--------------------------------------------------------------------
Output:
accuracy:               .1
approximate root:       49/32
decimal approximation:  1.53125

accuracy:               .01
approximate root:       793/512
decimal approximation:  1.548828125

accuracy:               .001
approximate root:       6347/4096
decimal approximation:  1.549560546875

accuracy:               .0001
approximate root:       50761/32768
decimal approximation:  1.549102783203125

accuracy:               .00001
approximate root:       812221/524288
decimal approximation:  1.5491886138916015625

accuracy:               .000001
approximate root:       6497789/4194304
decimal approximation:  1.5491936206817626953125

accuracy:               .0000001
approximate root:       51982303/33554432
decimal approximation:  1.54919335246086120605468755

accuracy:               .00000001
approximate root:       831716839/536870912
decimal approximation:  1.5491933356970548629760742187

accuracy:               .000000001
approximate root:       6653734721/4294967296
decimal approximation:  1.5491933377925306558609008789062

accuracy:               .0000000001
approximate root:       53229877789/34359738368
decimal approximation:  1.54919333840371109545230865478515625
--------------------------------------------------------------------
```

5.11 A Linear Fractional Cipher

A *cipher* consists of two algorithms: one to encode a message and one to decode it. In this section we use the fractional arithmetic capabilities of main to develop a new cipher. Our cipher is based on the notion of *linear fractional transformation*. These are functions of the form

$$y = \frac{ax+b}{cx+d}, \quad x \neq -d/c, \tag{5.1}$$

where a, b, c, d are fixed real constants and x is a real variable.

The encoding algorithm goes as follows: A message consists of capital letters A–Z, each of which has an *alpha value* given by the formula str(ord(letter)-65). For example, A has alpha value 0 and B has alpha value 1. An alpha value x of a letter is plugged into (5.1) to obtain a number y. This is subsequently converted into an integer fraction u/v. The message is the aggregate of the integer blocks u and v. All quantities in the program are in string form. Here is code for the algorithm.

```
def encryption(message,a,b,c,d):                        # string message
    coded_message = ''                                  # initialize
    for i in range(len(message)):
        x = str(ord(message[i]) - 65)                   # alpha value
        y = '(' + a + '*' + x + '+' + b +')/' \
            '(' + c + '*' + x + '+' + d + ')'           # formula for y
        y = main(y)[0]                                  # value of y
        u,v = y.split('/')                              # get num,den
        if v == '': v = '1'
        coded_message =  coded_message + u + ' ' + v + ' '
    L = len(coded_message)-1
    return coded_message[:L]                            # remove last space

-------------------------- Sample Run --------------------------
Input:
a = '15.23'; b = '42.72'; c = '63.91'; d = '27.45'      # arbitrary
message = 'RUNFORYOURLIVES'
coded_message = encryption(message,a,b,c,d)
print(coded_message)

Output:
30163 111392 34732 130565 24071 85828 11887 34700 25594 92219
30163 111392 13608 52043 25594 92219 34732 130565 30163 111392
21025 73046 968 3169 12085 45652 10364 28309 10562 39261
```

By way of explanation, the first pair 30163 111392 is the numerator and denominator of the output of main((17a+b)/(17c+d), where 17 is the alpha number of R, the first letter of the message.

5.11 A Linear Fractional Cipher

The decryption algorithm is the reverse process: it takes a coded message, the blocks generated by encryption, and converts it into the original message. This is possible because a linear fractional transformation has an inverse, obtained by solving for x in (5.1):

$$x = \frac{b - dy}{cy - a}. \tag{5.2}$$

Successive pairs of numbers in the coded message are formed into fractions y which are plugged into (5.2) to obtain the alpha value x. The corresponding character is obtained from the calculation chr(int(x)) + 65. For example, to decode the first pair 30163 111392 of the coded message, form the quotient $y = 30163/111392$ and plug this into (5.2) to obtain

$$\frac{b - d(30163/111392)}{c(30163/111392) - a} = 17,$$

and use chr(int(17)+65) to get back the letter R.

```
def decryption(coded_message,a,b,c,d):
    coded_message = coded_message.split(' ')   # split into blocks
    i = 0
    message = ''                                # initialize
    while i < len(coded_message)-1:
        u = coded_message[i]                    # first member of pair
        v = coded_message[i+1]                  # second member of pair
        y = '(' + '('+ u + ')/(' + v + ')' + ')'   # form fraction
        x = '(' + b + '-' + d + '*' + y +')/' \
            '(' + c + '*' + y + '-' + a + ')'   # formula for x
        x = ar.main(x)                          # value of x (alpha number)
        letter = chr(int(x) + 65)               # convert to its letter
        message = message + letter              # attach to message
        i += 2                                  # next pair
    return message

------------------------ Sample Run ---------------------------
Input:
a = '15.23'; b = '42.72'; c = '63.91'; d = '27.45'   # same as above
coded_message = \
30163 111392 34732 130565 24071 85828 11887 34700 25594 92219 \
30163 111392 13608 52043 25594 92219 34732 130565 30163 111392 \
21025 73046 968 3169 12085 45652 10364 28309 10562 39261
message = decryption(coded_message,a,b,c,d)
print(message)
-----------------------------------------------------------------
Output:
RUNFORYOURLIVES
-----------------------------------------------------------------
```

Polynomial Algebra

In this chapter we construct a module `PolyAlg.py` that performs algebraic operations on polynomials symbolically. The ultimate goal of the module is to convert an algebraic combination of polynomials into a single polynomial in standard form. The module is headed by the following import statements.

```
------------------------- PolyAlg.py --------------------------
import Number as nm
import Arithmetic as ar
import Tools as tl
import math as ma
---------------------------------------------------------------
```

6.1 Polynomial Operations

A *polynomial in x* is a mathematical expression of the form

$$a_m x^m + a_{m-1} x^{m-1} + \cdots + a_1 x + a_0, \tag{6.1}$$

where the a_k's are complex numbers, $a_m \neq 0$, and x is a variable. The integer m is called the *degree* of the polynomial. The expression $a_k x^k$ is called a *term* of the polynomial and a_k the *coefficient* of the term. The number a_0 is called the *constant term* and a_m the *leading coefficient*. The polynomial is said to be *monic* if the leading coefficient is one. A term of a polynomial is also called a *monomial*. We allow the constant a_0 to fall under this rubric, interpreting it as $a_0 x^0$. Thus the constant term, if there is one, may then be viewed as a coefficient.

Polynomials are added or subtracted by collecting terms with like powers. For multiplication, one must first expand. Here are some simple examples of these operations.

$$(5x^2 + 6x + 7) + (3x + 1) + = 5x^2 + 9x + 8$$
$$(5x^2 + 6x + 7) - (3x + 1) = 5x^2 + 3x + 6$$
$$(5x^2 + 6x + 7)(3x + 1) = 15x^3 + 23x^2 + 21x + 7$$

The general formula for multiplication is

$$(a_m x^m + a_{m-1} x^{m-1} + \cdots + a_1 x + a_0)(b_n x^n + b_{n-1} x^{n-1} + \cdots + b_1 x + b_0)$$
$$= c_p x^p + c_{p-1} x^{p-1} + \cdots + c_1 x + c_0,$$

where $p = m + n$ and c_k is the sum of all products $a_i b_j$ with $0 \leq i \leq m$, $0 \leq j \leq n$, and $i + j = k$. The special case $m = 0$ is simply multiplication of each term of the second polynomial by a_0, a process called *scalar multiplication*.

For programming purposes, a polynomial (6.1) is stored as a list P of its coefficients $[a_m, a_{m-1}, \ldots, a_1, a_0]$, so that $P[k] = a_{m-k}$. The items in the list are Gaussian rational numbers in the form of strings so that exact calculations via the module Arithmetic.py are possible. The algebraic operations on polynomials are carried out on the lists.

6.2 The Main Function

The function main(expr) takes as input an expression and returns the simplified polynomial and its list. For the output, one has the choice of a general fractional form, a monic form, or a form with Gaussian integer coefficients. The latter two forms generally have a compensating constant factor, which we shall call a *multiplier*. In the program the six outputs are labeled as follows: fpol, flist, mpol, mlist, ipol, and ilist. The multiplier is included as the first member of the lists mlist and ilist.

```
def main(expr):
    global var
    var = tl.get_var(expr)
    flist = expr2flist(expr)
    fpol = flist2pol(flist)              # general fractional form
    mlist = flist2mlist(flist)
    mpol = mlist2pol(mlist)              # monic form
    ilist = flist2ilist(flist)
```

```
            ipol = ilist2pol(ilist)                # integer coefficient form
            return fpol,flist,mpol,mlist,ipol,ilist

    ------------------------- Sample Run -------------------------
    Input:
    expr = '(3.1x +1-2.4i)^2 - 7x + 5/8i'
    print('fpol: ',fpol)
    print('flist:',flist)
    print('mpol: ',mpol)
    print('ipol: ',ipol)
    print('ilist',ilist)
    print('\n')
    print('fpol: ',ar.evaluate(fpol,varval,4)[0])
    print('mpol: ',ar.evaluate(mpol,varval,4)[0])
    print('ipol: ',ar.evaluate(ipol,varval,4)[0])
    -----------------------------------------------------------------

    Output:
    fpol:  (961/100)x^2+(-4/5-(372/25)i)x-119/25-(217/40)i
    flist: ['961/100', '-4/5-(372/25)i', '-119/25-(217/40)i']
    mpol:  (961/100)(x^2+(-80/961-(48/31)i)x-476/961-(35/62)i)
    ipol:  (1/200)(1922x^2+(-160-2976i)x-952-1085i)
    ilist: ['1/200', '1922', '-160-2976i', '-952-1085i']

    fpol:  ('179964661/1210000-(964769/11000)i', '148.7311-87.7063i')
    mpol:  ('179964661/1210000-(964769/11000)i', '148.7311-87.7063i')
    ipol:  ('179964661/1210000-(964769/11000)i', '148.7311-87.7063i')
    -----------------------------------------------------------------
```

6.3 Polynomial Operations in Python

For addition of polynomials it is convenient to have a function that prepends zeros to lists so as to make them the same length. For example, the polynomials

$$P(x) = 3x^2 + (2 + 3i)x + 1 \text{ and } Q(x) = 5x^3 + 7x,$$

are represented, respectively, by the lists

$$P = ['3','(2+3i)','1'], \quad Q = ['5','0','7','0'].$$

To give these lists equal length without changing the polynomials, a zero is prepended to P to obtain

$$['0','3','(2+3i)','1']$$

which represents the polynomial $P(x) = 0x^3 + 3x^2 + (2 + 3i)x + 1$. Addition is then carried out the lists term by term:

[0','3','(2+3i)','1'] + ['5','0','7','0'] = ['5','3','(9+3i)','1'],

which represents the polynomial $P(x)+Q(x) = 5x^3 + 3x^2 + (9+3i)x + 1$. Of course, prepending zeros violates the convention that the leading coefficient be nonzero, but in this context that is of no consequence. Here is the function that attaches the zeros:

```
def prependzeros(P,Q):
    numzeros = abs(len(P)-len(Q))
    Z = tl.zero_list(numzeros)
    if len(P) > len(Q): Q = Z+Q             # prepend zero list to Q
    if len(Q) > len(P): P = Z+P             # prepend zero list to P
    return P,Q
```

Since arithmetic operations may introduce leading zeros in a polynomial list, we need a function that removes these:

```
def remove_leading_zeros(P):
    if len(P) == 1:
        return P
    while len(P) > 1 and  P[0] == '0':
        P = P[1:]
    return P
```

---------------------------- Sample Run ----------------------------
```
Input:
print(remove_leading_zeros(['1','2','3'] ))
print(remove_leading_zeros(['0','0','1']))
print(remove_leading_zeros(['0','0']))
print(remove_leading_zeros(['0']))
```
```
Output:
['1', '2', '3']
['1']
['0']
['0']
```

The following functions add, subtract, and multiply polynomials. The inputs are polynomials P,Q in the form of lists, as described earlier.

```
def pol_sum(P,Q):
    P,Q = prependzeros(P,Q)
    S = []
    for i in range(len(P)):                              # add termwise
        s = ar.main('('+ P[i] +')+('+ Q[i] + ')')[0]
```

```
            S.append(s)
        S = remove_leading_zeros(S)
        return S

    def pol_scalar_prod(scalar,P):                   # returns scalar*P
        Q = []
        if isinstance(P,str): P = [P]
        for entry in P:                # multiply each item in P by scalar
            s = ar.main('('+ scalar +')*('+ entry + ')')[0]
            Q.append(s)
        return Q

    def pol_diff(P, Q):                              # returns  P-Q
        R = scalar_prod('-1',Q)
        return pol_sum(P,R)

    def pol_prod(P, Q):
        M = []
        L = len(P)+len(Q)-1
        for k in range(L):             # for each k calculate c_k
            ck = '0'                                 # reset
            for i in range(len(P)):
                for j in range(len(Q)):
                    if i+j == k:
                        ck = ar.main \
                        (ck + '+('+ P[i] +')('+ Q[j] + ')')[0]
            M.append(ck)
        return M

    def pol_power(P,n):
        Q = P
        if n > 1:
            for i in range(n-1):       # multiply P by itself n-1 times
                Q = pol_prod(P, Q)
        return Q

    def pol_quotient(P,q):                           # q a scalar
        z = ar.main('1/('+q+')')[0]
        return scalar_prod(z,P)
```

6.4 The Allocator

The following function scans the expression, assigning computational tasks to the functions in the preceding section. Scalars and variables are converted into their polynomial lists. The function is similar in logical structure to the eponymous function in Arithmetic.py.

```
def allocate_ops(expr,mode):
    global idx
    P = []
    while idx < len(expr):
        ch = expr[idx]
        if ch in '.0123456789i':
            start = idx
            r,idx = tl.extract_numeric(expr, start)
            P = [r]                             # list for numeric
        elif ch == var:                         # e.g. the 'x' in 'x^3'
            idx +=1                             # move from var to '^'
            exp,idx = tl.extract_exp(expr,idx)
            n = int(exp)
            P = ['1'] + tl.zero_list(n)         # list for x^n
        elif ch == '+':
            if mode > 0: break                  # wait for higher mode
            idx += 1
            Q = allocate_ops(expr,0)
            P = pol_sum(P,Q)
        elif ch == '-':
            if mode > 0: break                  # wait for higher mode
            idx += 1
            Q = allocate_ops(expr,1)
            P = pol_diff(P,Q)
        elif ch == '*':
            if mode > 1: break
            idx += 1
            Q = allocate_ops(expr,1)
            P = pol_prod(P,Q)
        elif ch == '^':
            idx += 1
            exp = allocate_ops(expr,2)[0]       # inside the list [exp]
            P = pol_power(P, int(exp))
        elif ch == '/':
            if mode > 1: break
            idx += 1
            Q = allocate_ops(expr,1)            # denominator, a scalar
            q = Q[0]
            P = pol_quotient(P,q)               # divide by scalar
        elif ch == '(':
            start = idx
            paren_expr,end = tl.extract_paren(expr,start)
            if not var in paren_expr:
                r = ar.complex_calc(paren_expr)[0]
                P = [r]
                idx=end
            else:
                idx+=1
                P = allocate_ops(expr,0)
                idx+=1
        elif ch == ')': break
    return P
```

6.5 Generating the Polynomial Lists

The function `expr2flist(expr)` prepares the expression `expr`, making it suitable for digestion by the function `allocate_ops`, and then applies the latter function to obtain the general fractional coefficient list `flist`.

```
def expr2flist(expr):
    global idx, var
    expr = tl.attach_missing_exp(expr,var)
    expr = tl.insert_asterisks(expr,var)
    idx = 0                                  # start of expression
    flist = allocate_ops(expr,0)             # fractional coefficients
    return flist
```

The function `flist2mlist(flist)` takes the list generated by `expr2flist` and returns the corresponding monic form `mlist`. It calculates the reciprocal of the leading coefficient in `list`. Each member of `flist` except the first is multiplied by the reciprocal.

```
def flist2mlist(flist):
    if len(flist) <= 1:
        return flist                         # trivial list
    mlist = []
    multiplier = flist[0]                    # leading coefficient of flist
    # invert leading coefficient
    reciprocal = ar.main( '1/('+ multiplier + ')')[0]
      # multiply each coefficient except the first by reciprocal:
    for c in flist[1:]:
        z = ar.main('('+ reciprocal + )('+ c +')')[0]
        mlist.append(z)
    mlist = ['1'] + mlist                    # make monic
    mlist = [multiplier] + mlist             # prepend compensating factor
    return mlist
```

The function `flist2ilist(flist)` takes the list generated by `expr2flist` and returns the corresponding integer coefficient form `ilist`. To do this it multiplies the members of `flist` by the least common multiple of their denominators. The compensating multiplier, the reciprocal of the lcm, is then attached to the result, producing `ilist`. The function `get_denoms(flist)` runs through `flist` gathering the denominators.

```
def flist2ilist(flist):
    ilist = []
    denoms = get_denoms(flist)
    lcm = nm.listlcm(denoms)                 # get lcm of denoms
```

```
        for item in flist:              # multiply each item by the lcm
            prod = ar.main(str(lcm) + '*(' + item + ')')[0]
            ilist.append(prod)
        multiplier = '1/'+ str(lcm)      # compensatory factor
        multiplier = ar.main(multiplier)[0]              # reduce it
        ilist = [multiplier] + ilist    # prepend multiplier to ilist
        return ilist

    def get_denoms(flist):
        denoms = [1]                                          # default
        for i in range(len(flist)):
            coeff = flist[i]
            re = ar.real(coeff)
            im = ar.imag(coeff)
            if '/' in re:                # get denominator of real part
                den = int(re.split('/')[1])
                denoms.append(den)
            if '/' in im:                # get denominator of imag part
                den = int(im.split('/')[1])
                denoms.append(den)
        return list(set(denoms))                   # remove duplicates

    --------------------------- Sample Run ---------------------------
    Input:
    flist = ['1/12','1/5','1/4','1/3']
    print(get_denoms(flist))
    ------------------------------------------------------------------
    Output:
    [1, 3, 4, 5, 12]
    ------------------------------------------------------------------
```

6.6 Converting Lists to Polynomials

The functions in this section produce formatted polynomials from the lists generated by the functions in the preceding section. The first, flist2pol(flist,var), attaches the coefficients to powers of the variable (denoted by the generic letter 'x' in the comments).

```
    ------------------------------------------------------------------
    def flist2pol(flist):
        P = attach_parens(flist)   # enclose fractions and complex no.s
        pol = ''                              # for polynomial string
        L = len(P)
        # attach coeffs to var
        for exp in range(L):                          # exponent of term
            v = var
            if exp == 0: v = ''                           # omit x^0
            coeff = P[L-exp-1]                # coefficient of x^exp
```

6.6 Converting Lists to Polynomials

```
                if coeff == '0':
                    continue                    # skip zero coeff
                if coeff == '1' and exp != 0:
                    coeff = ''      # 1 superfluous
                if coeff == '-1' and exp != 0:
                    coeff = '-'
                term = coeff + v
                if exp > 1:                     # omit power x^exp for exp = 1
                    term = term + '^' + str(exp)
                pol = term + '+' + pol
        if pol == '': return '0'
        pol =  pol[:len(pol) - 1]               # remove extra '+'
        pol = pol.replace('( - ', '(-')         # make pretty
        pol = tl.fix_signs(pol)
        return pol

    def attach_parens(pol_list):
        L = len(pol_list)
        if L <= 1: return pol_list
        P = []
        for k in range(L-1):     # attach parens to all but constant term
            coeff = pol_list[k]
            #if '/' in coeff or '+' in coeff or '-' in coeff:
            coeff = tl.add_parens(coeff)
            coeff = tl.fix_signs(coeff)
            P.append(coeff)
        P.append(pol_list[L-1])                 # pick up constant term
        return P

    -------------------------- Sample Run --------------------------
    Input:
    flist = ['-1/2','3', '2i', '5-7i', '2-4i']
    print(flist2pol(flist,'x'))
    ----------------------------------------------------------------
    Output:
    (-1/2)x^4+3x^3+2ix^2+(5-7i)x+2-4i
    ----------------------------------------------------------------
```

For the monic and integer forms of the polynomial we need to attach the multiplier, which may require parentheses to be placed around it and/or the polynomial. Polynomials with only one term require no parentheses. The following function detects this state of affairs. It takes a list of coefficients and returns True if the list has only one element or if it has only one nonzero member.

```
    ----------------------------------------------------------------
    def is_single_term(pol_list):
        L = len(pol_list)
        if L == 1:
            return True                 # polynomial has only one term
        numzeros = 0
        for item in pol_list:           # calculates number n of zeros in list
```

```
            if item == '0':
                numzeros +=1
        return numzeros == L-1                  # ['1','0','0']: L -1 = 2
```

The function attach_multiplier(pol_list,pol,factor) returns the monic or integer coefficient polynomial with the multiplier attached and with suitable parentheses.

```
def attach_multiplier(pol_list,pol,multiplier):
    if multiplier == '1': return pol
    if ar.is_complex(multiplier) or ar.is_frac(multiplier):
        multiplier = '(' + multiplier + ')'
    if not is_single_term(pol_list):
        pol = '(' + pol + ')'
    return multiplier+pol
```

The next two functions return monic and integer coefficient polynomials.

```
def mlist2pol(mlist):
    if len(mlist) == 1:
    return mlist[0]
    multiplier = mlist[0]
    P = mlist[1:]                                # list without multiplier
    pol = flist2pol(P)
    return attach_multiplier(P,pol,multiplier)

def ilist2pol(ilist):
    multiplier = ilist[0]
    P = ilist[1:]
    pol = flist2pol(P)
    return attach_multiplier(P,pol,multiplier)
```

6.7 The Modular Case

In this section we consider the case where the polynomials have integer coefficients modulo some integer $m > 1$. Since there is no division involved, the same calculation functions work for this case. The only difference is that the coefficients of the final polynomial need to be replaced by their remainders modulo m. The function poly_mod(expr,v,m) takes an expression in the variable var and with integer coefficients and returns the reduced polynomial with coefficients modulo m

```
def strmod(a,m):                                         # string 'a'
    int_a = int(a)
    b = (int_a) % m
    return str(b)                        # modulus in string form

def integerlist2modlist(L,m):
    # converts each int in list to mod m number
    modlist = []
    for k in range(len(L)):
        modm = strmod(L[k],m)
        modlist = modlist + [modm]
    return remove_leading_zeros(modlist)

def poly_mod(expr,m):
    fpol,flist = main(expr)[0:2]                # integer coeff's
    modlist = integerlist2modlist(flist,m)
    modpol = flist2pol(modlist)
    return modpol, modlist

------------------------- Sample Run -------------------------
Input:
expr = '(3x-57)^12'
for m in range(2,14):
    print('mod',m,' ', poly_mod(expr,'x',m)[0])
--------------------------------------------------------------
Output:
mod 2    x^12+x^8+x^4+1
mod 3    0
mod 4    x^12+2x^10+3x^8+3x^4+2x^2+1
mod 5    x^12+2x^11+x^10+2x^7+4x^6+2x^5+x^2+2x+1
mod 6    3x^12+3x^8+3x^4+3
mod 7    x^12+3x^11+5x^10+3x^9+3x^8+4x^7+2x^5+6x^4+3x^3+6x^2+6x+1
mod 8    x^12+4x^11+2x^10+4x^9+7x^8+4x^6+7x^4+4x^3+2x^2+4x+1
mod 9    0
mod 10   x^12+2x^11+6x^10+5x^8+2x^7+4x^6+2x^5+5x^4+6x^2+2x+1
mod 11   9x^12+5x^11+5x+4
mod 12   9x^12+6x^10+3x^8+3x^4+6x^2+9
mod 13   x^12+6x^11+10x^10+8x^9+9x^8+2x^7+12x^6+7x^5+3x^4+5x^3\
         +4x^2+11x+1
--------------------------------------------------------------
```

6.8 Application: Completing the Square

The function complete_square(quad) takes a quadratic $ax^2 + bx + c$ and returns its completed square $a(x + (1/2)(b/a))^2 + c - a((1/2)(b/a))^2$.

```
def complete_square(quad):
    # a = quad[0], b = quad[1], c = quad[2]
    coeffs = main(quad)[1]                              # flist
    a = coeffs[0]
    b = coeffs[1]
    c = coeffs[2]
    d = ar.main('(1/2)('+ b +'/' + a + ')')[0]          # (1/2)(b/a)
    e = ar.main(a+'('+ d + ')^2')[0]                    # a((1/2)(b/a))^2
    f = ar.main(c+ '-' + e)[0]                          # c - a((1/2)(b/a))^2
    g = main('(x' + '+' + d + ')')[0]                   # x + (1/2)(b/a)
    s = tl.fix_signs(a + '('+ g +')^2' + '+' + f)
    return s
#-------------------------- Sample Run --------------------------
Input:
q = '3x^2 - 8x + 5'
print(complete_square(q))
--------------------------------------------------------------
Output:
3(x-4/3)^2-1/3
```

6.9 Application: Lagrange Interpolation

An *interpolation function* for a set of data points

$$(x_1, y_1), (x_2, y_2), \ldots, (x_n, y_n), \quad x_j \neq x_k \text{ for } j \neq k, \tag{6.2}$$

is a function $f(x)$ of a continuous variable x such that $f(x_k) = y_k$ for all k. The underlying assumption here is that f generates hypothetical data values $(x, y = f(x))$ for values of x that are not necessarily the observed values. Such a function usually has a relatively simple form so that the *interpolated* values y are easily calculated. Generally, these are only approximations to the actual data generated by the underlying process but may be sufficiently accurate to be useful. In this section we consider one of the simplest and most common forms of interpolation and implement the formula in Python.

The *Lagrange interpolation polynomial* for the data points in (6.2) is the degree n polynomial

$$P(x) = \sum_{j=1}^{n} P_j(x),$$

where

$$P_j(x) = y_j L_j(x) \text{ and } L_j(x) = \prod_{\substack{k=1 \\ k \neq j}}^{n} \frac{x - x_k}{x_j - x_k}$$

6.9 Application: Lagrange Interpolation

The large pi symbol stands for product and is used like the large sigma in summation. Since

$$L_j(x_j) = 1 \text{ and } L_j(x_i) = 0 \text{ for } i \neq j,$$

$P(x)$ has the desired interpolation property $P(x_k) = y_k$ for all k.

The function `lagrange_interp(data)` takes data and returns the expanded version of the interpolation function. For convenience we use the function `data2lists(data)` to convert the entered data (6.2) to a table.

```
def data2lists(data):
    # e.g. data =  '(1.7,3.2),(2.2,5.4),(-2.98,.76)'
    data = data.replace(' ','')
    data = data.replace('),(',';')    # (1.7,3.2; 2.2,5.4; -2.98,.76)
    data = data.replace(')','')       # (1.7,3.2; 2.2,5.4; -2.98,.76
    data = data.replace('(','')       # 1.7,3.2; 2.2,5.4; -2.98,.76'
    tab = tl.string2table(data)       # convert data to tabular form
    return tab     # [['1.7','3.2'], ['2.2','5.4'], ['-2.98','.76']]

def lagrange_interp(data):
    sum = '0'
    data = data2lists(data)
    for j in range(len(data)):
        denprod = '1'
        numprod = '1'
        # get products of numerators and denominators in formula:
        for k in range(len(data)):
            if k != j:
                num = '(x' '-(' + data[k][0] + '))'
                numprod = num + '('+ numprod + ')'
                numprod = main(numprod)[0]
                den = '('+ data[j][0] +'-('+ data[k][0] +')'+')'
                denprod = ar.main(den +'('+ denprod +')')[0]
        Lj = '(' + numprod + ')/(' + denprod + ')'
        Pj = main('('+ data[j][1] +')('+ Lj +')')[0]
        sum = main(sum +'+'+ Pj)[0]
    return main(sum)[0]
```
```
Input:
data =  '(1.7,3.2),(2.2,5.4),(-2.98,.76)'
pol = lagrange_interp(data)
print('Lagrange polynomial:\n', pol)
```
```
Output:
Lagrange polynomial:
(22690/30303)x^2 + (5749/3885)x - 1120906/757575
```

6.10 Application: Polynomial Calculus

In this section we use some of the primary functions developed thus far to construct functions that differentiate and integrate polynomials. We treat differentiation first.

First Derivative of a Polynomial

The function deriv(P) returns the first derivative of a polynomial. It does so by multiplying the members of the list of P by their index, the exponent, counted from the right. For example, the polynomial

$$P(x) = 5x^3 + 7x^2 + 9x + 11$$

with list PL=[5,7,9,11] has derivative

$$Q(x) = 15x^2 + 14x + 9$$

with list QL=[15,14,9]. If L denotes the length of PL, then QL may be seen as the list

$$[5*(L-1), 7*(L-2), 9*(L-3), 11*(L-4)].$$

Here is the code:

```
def deriv(P):
    flist = main(P)[1]
    L = len(flist)
    dflist = []                     # for the flist of the derivative
    for k in range(L-1):
        exp = str(L-1-k)
        coeff = flist[k]
        dcoeff = ar.main( exp + '(' + coeff + ')')[0]
        dflist = dflist + [dcoeff]
    return flist2pol(dflist)

------------------------- Sample Run -------------------------
Input:
P = '(3/2)x^12 + 2.004 x^10 - 5ix^8 + 6x^4 + 7x^2 + (11-5i)x + 7.2'
print(deriv(P))

Output:
18x^11+(501/25)x^9-40ix^7+24x^3+14x+11-5i
```

6.10 Application: Polynomial Calculus

Tangent Line

We can use deriv to find the equation of the line tangent to a polynomial P at a point $(a, P(a))$, namely, $y = P(a) + P'(a)(x - a)$.

```
def tangent_line(P,a):
    Q = deriv(P)
    b = ar.evaluate(P,a,'')[0]    # = P(a)
    m = ar.evaluate(Q,a,'')[0]    # P'(a) = slope
    tan_eqn = b + ' +('+ m +')('+ 'x' + '-' + a +')'
    tan_eqn = tl.fix_signs(tan_eqn)
    return 'y = '+ tan_eqn
```

```
------------------------ Sample Run ------------------------
Input:
P = 'x^4 + 5x^3 - 3x^2 + 7x - 9'
a = '-3/4'
print(tangent_line(P,a))

Output:
y = -4539/256+(73/4)(x+3/4)
```

Higher Order Derivatives

The function dderiv(P,n) finds the *nth* derivative of a polynomial P.

```
def dderiv(P,n):
    d = P
    for k in range(n):
        d = deriv(d)            # derivative of previous derivative
    return d
```

```
------------------------ Sample Run ------------------------
Input:
P = '(2.4+3.1i)x^10 - 5ix^8 + 6x^4 + 7x^2 + 11'
for n in range(12):
    print(n,' ', dderiv(P,n))

Output:
0    (2.4+3.1i)x^10 - 5ix^8 + 6x^4 + 7x^2 + 11
1    (24+31i)x^9+(-40i)x^7+24x^3+14x
2    (216+279i)x^8+(-280i)x^6+72x^2+14
3    (1728+2232i)x^7+(-1680i)x^5+144x
4    (12096+15624i)x^6+(-8400i)x^4+144
5    (72576+93744i)x^5+(-33600i)x^3
6    (362880+468720i)x^4+(-100800i)x^2
7    (1451520+1874880i)x^3+(-201600i)x
8    (4354560+5624640i)x^2-201600i
```

```
 9         (8709120+11249280i)x
10         8709120+11249280i
11         0
```

Taylor Series

The *Taylor series expansion* of an infinitely differentiable function $f(x)$ about a point a is an infinite series in x that, under suitable conditions, converges to $f(x)$:

$$f(x) = \sum_{k=0}^{\infty} \frac{f^{(k)}(a)}{k!}(x-a)^k.$$

Here $f^{(k)}$ denotes the kth derivative of f, where, by convention, $f^{(0)} = f$. The partial sums

$$T_n(x) = \sum_{k=0}^{n} \frac{f^{(k)}(a)}{k!}(x-a)^k$$

are called *Taylor polynomials* and are useful for approximating f and its integrals.

For polynomials f the series is finite since for k greater than the degree d of f, $f^{(k)} = 0$. The Taylor series then coincides with T_n for $n > d$. The following function returns the Taylor series for polynomials for the case n = degree of f. The run includes a check.

```
def taylor_series(P,a):
    var = tl.get_var(P)
    PL = main(P)[1]                                  # list for P
    t = ''                                           # initialize series
        # calculate derivatives up to order degree P = len(PL)-1:
    for n in range(len(PL)):                         # generate Taylor series terms
        DP = dderiv(P,n)
        c = '('+ ar.evaluate(DP,a,'')[0]+')/('+str(ma.factorial(n))+')'
        c = ar.main(c)[0]                            # coefficient of term
        if c == '1':  c = ''                         # remove extraneous '1'
        if c == '-1': c = '-'
        c = tl.add_parens(c)
        if n == 1:
            t = c + '(' + var + '-' + a + ')' + '+' + t
        elif n == 0:
            t = c + t
        else:
            t = c + '(' + var + '-' + a + ')^' + str(n) + '+' + t
    t = t[:len(t)]
    return tl.fix_signs(t)

------------------------- Sample Run ---------------------------
Input:
P = 'x^4 - x^3 + x^2 + x + 1'; a = '1.1'
T = taylor_series(P,a)
```

```
print(T,'\n')
# check:
print(main(T)[0])
```

```
Output:
(x-1.1)^4+(17/5)(x-1.1)^3+(124/25)(x-1.1)^2+(2447/500)(x-1.1) \
+(34431/10000)

'x^4 - x^3 + x^2 + x + 1'
```

Single Integration of a Polynomial

Recall that an indefinite integral (antiderivative) of a function is determined only up to a constant c, the so-called constant of integration. The function indef_integral(P) below returns the indefinite integral of a polynomial P for the case $c = 0$. As with derivatives, the function works on the list for P. Specifically it divides members of the list by one plus their index counted from the right. For example, the polynomial

$$P(x) = 5x^3 + 7x^2 + 11$$

with list PL=[5,7,0,11] has integral

$$Q(x) = (5/4)x^4 + (7/3)x^3 + 11x + 0$$

with list QL=[5/4,7/3,0,11,0]. If L denotes the length of PL then QL may be seen as

$$[5/L, 7/(L-1), 0/(L-2), 11/L-3, 0].$$

The following code below implements this procedure. The sample run illustrates the familiar fact that the operations of indefinite integration and differentiation are, up to a constant, inverses of each other.

```
def indef_int(P):
    flist = main(P)[1]                          # polynomial list
    L = len(flist)
    int_flist = []
    for k in range(0,L):
        exp = str(L-k)
        coeff = flist[k]
        int_coeff = ar.main('('+ coeff +')/('+ exp +')')[0]
        int_flist = int_flist + [int_coeff]
    int_flist = int_flist + ['0']
    return flist2pol(int_flist)
```

```
-------------------------- Sample Run --------------------------
Input:
P = '(2.4+3.1i)x^10 - 5ix^8 + 6x^4 + 7x^2 + 13'
I = indef_int(P)
print('indefinite integral of P:')
print(I,'\n')

D = deriv(I)
print('derivative of indefinite integral of P:','\n')
print(D,'\n')

D = deriv(P)
I = indef_int(D)
print('indefinite integral of derivative of P:')
print(I)
----------------------------------------------------------------
Output:
indefinite integral of P:
(12/55+(31/110)i)x^11+(-5/9)ix^9+(6/5)x^5+(7/3)x^3+13x

derivative of indefinite integral of P:
(12/5+(31/10)i)x^10-5ix^8+6x^4+7x^2+13

indefinite integral of derivative of P:
(12/5+(31/10)i)x^10-5ix^8+6x^4+7x^2
----------------------------------------------------------------
```

To find a definite integral of P we first get an indefinite integral Q and then apply the fundamental theorem of calculus,

$$\int_a^b P(x)\,dx = Q(b) - Q(a).$$

```
----------------------------------------------------------------
def_integral(P,a,b):
    var = tl.get_var(P)
    I = indef_integral(P)
    A = ar.evaluate(I,a,'')[0];
    B = ar.evaluate(I,b,'')[0]
    return ar.main(B+'-('+A+')')[0]    # fundamental theorem of calc.

-------------------------- Sample Run --------------------------
Input:
P = '(3/2)x^12 + 2.4 x^10 - 5ix^8 + 6x^4 + 7x^2 + 11'
a = '1'; b = '3'
print(def_integral(P,a,b))
----------------------------------------------------------------
Output:
478299251/2145-(98410/9)i
```

6.10 Application: Polynomial Calculus

The function indef_integral_with_condition finds the indefinite integral I of a polynomial P subject to a condition of the form $I(s) = t$. This enables the determination of the constant of integration. For example, the general indefinite integral of the polynomial $P(x) = x^2 + 3x + 5$ is

$$I(x) = (1/3)x^3 + (3/2)x^2 + 5x + c.$$

If we require that $I(1) = 7$ we obtain the equation $7 = 1/3 + 3/2 + 5 + c$. Thus $c = 2 - 1/3 - 3/2 = 1/6$ and so

$$I(x) = (1/3)x^3 + (3/2)x^2 + 5x + 1/6.$$

```
def indef_integral_with_condition(P,s,t):    # I(s) = t
    I = indef_int(P)
    B = ar.evaluate(I,s,'')[0]
    c = t + '-('+ B +')'  # solve for c
    return main(I + '+' + c)[0]
```

---------------------------- Sample Run ----------------------------
```
Input:
P = 'x^3 - 7x^2 + 9'
s = '1.1'; t = '2.7'
I = indef_integral_with_condition(P,s,t)
print('I = ',I)
print(t,ar.evaluate(I,s,'')[0])              # check if equal
```

```
Output:
I = (1/4)x^4+(-7/3)x^3+9x-535243/120000
2.7 27/10                                    # I(1.1) = 2.7
```

Iterated Polynomial Integration

The function indef_integral_with_conditions takes a polynomial P and a list of n conditions and applies the function indef_integral_with_condition n times. For example, suppose that we wish to integrate $P(x)$ twice such that the first integration $I_1(x)$ is subject to the condition $I_1(s_1) = t_1$ and the second is subject to $I_2(s_2) = t_2$. The function indef_integral_with_condition enables us to find the constant of integration and therefore the complete determination of I_1. Integrating the latter we obtain $I_2(x)$. Using the above function again we can find the constant of integration and therefore completely determine I_2. For a concrete example, let $A(t)$ denote the acceleration of a particle at time t. If the initial velocity $V(0)$ and the initial position $S(0)$ of the particle are known, then the particle's position $S(t)$ for all time t may be determined by two integrations, one for $A(t)$ to find $V(t)$, using the initial condition for V to evaluate the constant of integration, the other for $V(t)$ to find $S(t)$, using the initial condition for S.

```
def indef_integral_with_conditions(P,C)
    C = tl.string2table(C)    # e.g. 's1,t1;s2,t2'-->[[s1,t1],[s2,t2]]
    I = P
    for item in C:
        s = item[0]; t = item[1]
        I = indef_integral_with_condition(I,s,t)
        print('I = ',I)                          # check progress
    return main(I,'f')[0]                        # format I
```

```
-------------------------- Sample Run --------------------------
Input:
C = '1,2; 3,7; 5,9'                              # three integrations
P = 'x^2-3x+ 7'
I = indef_integral_with_conditions(P,C)
-----------------------------------------------------------------
Output:
I = (1/3)x^3+(-3/2)x^2+7x-23/6                   # I(1) = 2
I = (1/12)x^4+(-1/2)x^3+(7/2)x^2+(-23/6)x-25/4   # I(3) = 7
I = (1/60)x^5+(-1/8)x^4+(7/6)x^3+(-23/12)x^2 \
    +(-25/4)x-253/8                              # I(5) = 9
-----------------------------------------------------------------
```

6.11 Application: Special Polynomials

The polynomials described in this section have numerous applications in both pure and applied mathematics. These applications are beyond the scope of the present text. Our goal here is simply to develop Python and functions that generate these polynomials. While there are many ways to do this, the one that is most suited to the spirit of the book is simple recursion. We shall use this method throughout. Each of the functions takes a positive integer n and returns a polynomial in the variable x and with degree n. The functions are defined by recurrence relations which define the recursive aspect of the function. Specifically, the relations express a polynomial of degree n as a function of similar polynomials of lesser degree.

Chebyshev Polynomials

The recurrence relation for these polynomials is

$$P_0 = 1, \quad P_1 = x,$$
$$P_n = 2xP(n-1) - P(n-2), \; n > 2.$$

6.11 Application: Special Polynomials

The function ChebyshevPol(n) implements the recurrence.

```
def ChebyshevPol(n):
    if n == 0: return '1'
    if n == 1: return 'x'
    A = ChebyshevPol(n-1)
    B = ChebyshevPol(n-2)
    pol = '(2x)('+ A + ')-('+ B + ')'
    return main(pol)[4]
```

```
------------------------- Sample Run -------------------------
Input:
for n in range(11): print(ChebyshevPol(n))
--------------------------------------------------------------
Output:
1
x
2x^2 - 1
4x^3 - 3x
8x^4 - 8x^2 + 1
16x^5 - 20x^3 + 5x
32x^6 - 48x^4 + 18x^2 - 1
64x^7 - 112x^5 + 56x^3 - 7x
128x^8 - 256x^6 + 160x^4 - 32x^2 + 1
256x^9 - 576x^7 + 432x^5 - 120x^3 + 9x
512x^10 - 1280x^8 + 1120x^6 - 400x^4 + 50x^2 - 1
--------------------------------------------------------------
```

Legendre Polynomials

The recurrence relation for these polynomials is

$$P_0 = 1, \quad P_1 = x,$$
$$P_n = (2 - 1/n)xP(n-1) + (1/n - 1)P(n-2), \quad n > 2.$$

The function LegendrePol(n) implements the relation.

```
def LegendrePol(n):
    if n == 0: return '1'
    if n == 1: return 'x'
    A = LegendrePol(n-1)              # get the 2 previous pols
    B = LegendrePol(n-2)
    pol = '(2-1/'+ str(n) +')x('+ A +')+(1/'+str(n)+'-1)('+ B +')'
    return main(pol)[4]
```

```
------------------------- Sample Run -------------------------
Input:
for n in range(11): print(LegendrePol(n))
```

```
Output:
1
x
(1/2)(3x^2 - 1)
(1/2)(5x^3 - 3x)
(1/8)(35x^4 - 30x^2 + 3)
(1/8)(63x^5 - 70x^3 + 15x)
(1/16)(231x^6 - 315x^4 + 105x^2 - 5)
(1/16)(429x^7 - 693x^5 + 315x^3 - 35x)
(1/128)(6435x^8 - 12012x^6 + 6930x^4 - 1260x^2 + 35)
(1/128)(12155x^9 - 25740x^7 + 18018x^5 - 4620x^3 + 315x)
(1/256)(46189x^10 - 109395x^8 + 90090x^6 - 30030x^4 + 3465x^2 - 63)
```

Laguerre Polynomials

The recurrence relation for these polynomials is

$$P_0 = 1, \quad P_1 = 1 - x,$$
$$P_n = (2 - 1/n - x/n)P(n-1) + (1/n - 1)P(n-2), \quad n > 2.$$

```
def LagurrePol(n):
    if n == 0: return '1'
    if n == 1: return '1-x'
    A = LagurrePol(n-1)
    B = LagurrePol(n-2)
    pol = '(2-1/'+str(n)+ '-x/'+ str(n) +')('+ A +')+ \
          (1/'+str(n)+'-1)('+ B +')'
    return main(pol)[4]
```

---------------------------- Sample Run ----------------------------
```
Input:
for n in range(11): print(LagurrePol(n))
```
```
Output:
1
1-x
(1/2)(x^2 - 4x + 2)
(1/6)(- x^3 + 9x^2 - 18x + 6)
(1/24)(x^4 - 16x^3 + 72x^2 - 96x + 24)
(1/120)(- x^5 + 25x^4 - 200x^3 + 600x^2 - 600x + 120)
(1/720)(x^6 - 36x^5 + 450x^4 - 2400x^3 + 5400x^2 - 4320x + 720)
(1/5040)(- x^7 + 49x^6 - 882x^5 + 7350x^4 - 29400x^3 + 52920x^2 \\
         - 35280x + 5040)
(1/40320)(x^8 - 64x^7 + 1568x^6 - 18816x^5 + 117600x^4 \\
          - 376320x^3 + 564480x^2 - 322560x + 40320)
```

```
(1/362880)(- x^9 + 81x^8 - 2592x^7 + 42336x^6 - 381024x^5 \\
         + 1905120x^4 - 5080320x^3 + 6531840x^2 - 3265920x \\
         + 362880)
(1/3628800)(x^10 - 100x^9 + 4050x^8 - 86400x^7 + 1058400x^6 \\
          - 7620480x^5 + 31752000x^4 - 72576000x^3 + 81648000x^2 \\
          - 36288000x + 3628800)
```

Hermite Polynomials

The recurrence relation for these polynomials is

$$P_0 = 1, \quad P_1 = 1 - x,$$
$$P_{n+1} = x P_n - P_n'$$

```
def HermitePol(n):
    if n == 0: return '1'
    A = HermitePol(n-1)
    B = deriv(A)
    pol = 'x('+ A + ')-('+ B + ')'
    return main(pol,'i')[0]
```

---------- Sample Run ----------
Input:
for n in range(11): print(HermitePol(n))

Output:
```
1
x
x^2-1
x^3+(-3)x
x^4+(-6)x^2+3
x^5+(-10)x^3+15x
x^6+(-15)x^4+45x^2-15
x^7+(-21)x^5+105x^3+(-105)x
x^8+(-28)x^6+210x^4+(-420)x^2+105
x^9+(-36)x^7+378x^5+(-1260)x^3+945x
x^10+(-45)x^8+630x^6+(-3150)x^4+4725x^2-945
```

7 Polynomial Divisibility and Roots

A nonconstant polynomial B is said to *divide* a polynomial A if there exists a nonconstant polynomial Q such that $A(x) = B(x)Q(x)$. The polynomials B and Q are then called *divisors* or *factors* of A. For example, $3x + 2$ and $(4x - 1)$ divide $12x^2 + 5x - 2$ since $12x^2 + 5x - 2 = (3x + 2)(4x - 1)$. We exclude constant polynomials from these definitions to avoid trivialities. In this chapter we construct a module centered around these ideas. Specifically, we construct polynomial versions of the division algorithm and the extended greatest common divisor, both of which were discussed in Chap. 4. All polynomials have coefficients that are Gaussian rational numbers or integers modulo a positive integer. The module is headed by the import statements

```
------------------------- PolyDiv.py --------------------------
import PolyAlg as pl
import Arithmetic as ar
import Tools as tl
import Number as nm
import math as ma
---------------------------------------------------------------
```

7.1 Division Algorithm for Polynomials

The reader will recall that the division algorithm for integers a and b asserts the existence of unique integers q and r such that

$$a = qb + r, \quad 0 \leq r < |b|.$$

There is an analogous theorem for polynomials. It asserts that for given polynomials A and B there exist polynomials Q and R (unique up to a constant) such that

$$A = QB + R, \quad \deg(R) < \deg(B).$$

As in the case of integers, A is called the *dividend*, B the *divisor*, Q the *quotient*, and R the *remainder*. For example, for $A = x^5 - 2x^3 + 7x^2 - 5x + 11$ and $B = 2x^2 + 1$, one has

$$Q(x) = (1/2)x^2 - (5/4)x + 7/2 \text{ and } R(x) = -(15/4)x + 15/2,$$

as the reader may check.

The polynomials Q and R may be obtained using long division of A by B. The following is an example of the long division algorithm organized in a manner that follows the coded implementation below. We have included all coefficients, zeros as well, as these form the lists AL, BL, and CL in the code.

$$
\begin{array}{r|lr}
 & (1/2)\,x^3 \;-\; (5/4)\,x \;+\; (7/2) & Q \\
\text{BL } 2x^2 + 0x + 1 \Big) & 1x^5 + 0x^4 \;-\; 2x^3 + 7x^2 \;-\; 5x \;+11 & \text{AL} \\
 & 1x^5 + 0x^4 + (1/2)\,x^3 + 0x^2 + 0x \;+\;0 & \text{CL} \\ \hline
 & 0x^5 + 0x^4 - (5/2)\,x^3 + 7x^2 \;-\; 5x \;+11 & \text{AL} \\
 & -(5/2)\,x^3 + 0x^2 - (5/4)\,x\;+\;0 & \text{CL} \\ \hline
 & 7x^2 - (15/4)\,x + 11 & \text{AL} \\
 & 7x^2 - \qquad\qquad 0\,x + 7/2 & \text{CL} \\ \hline
 & -(15/4)\,x + 15/2 & \text{RL}
\end{array}
$$

The function div_alg(A,B) implements the algorithm. It takes polynomials A, B and returns the quotient Q and remainder R.

```
def div_alg(A,B):                          # polynomial strings
    var = tl.get_var(A)
    pl.var = var                           # global for PolyAlg
    AL = pl.main(A)[1]                     # polynomial list for A
    BL = pl.main(B)[1]                     # polynomial list for B
    degree_A = len(AL) - 1
    degree_B = len(BL) - 1
    if degree_A < degree_B:                # trivial case
        Q = '0';
        R = A;
        return Q, R
    if degree_B == 0 and B != '0':
        Q = pl.main('('+A+')/('+B+')')[0]  # divide by constant B
        R = '0'
        return Q, R
    Q = ''                                 # initialize quotient string
```

7.2 Extended GCD for Polynomials

```
    while True:
        degree = len(AL) - len(BL)
        q = '('+ AL[0] +')/('+ BL[0] +')'      # q = AL[0]/BL[0]
        q = ar.main(q)[0]
        # attach power to q:
        Q = Q + '+('+ '('+ q +')' + var + '^' + str(degree) +')'
        Q,QL = pl.main(Q)[0:2]
        CL = pl.scalar_prod(q,BL)
        # attach 0's to DL to match degree of AL
        CL = CL + tl.zero_list(len(AL) - len(BL)) len(BL))]
        AL = pl.pol_diff(AL,CL)
        if len(AL)< len(BL):
            break                              # done
    RL = AL
    R = pl.flist2pol(RL)                       # convert to polynomial
    return Q, R

------------------------- Sample Run -------------------------
Input:
A =  6x^7 + (2/5+3i)x^3 + (5.3/2)x^2 - 3.4x + 2
B =  2x^2+x+4/i
Q, R = div_alg(A,B)
print('Q =',Q)
print('R =',R)
--------------------------------------------------------------
Output:
Q = 3x^5+(-3/2)x^4+(3/4+6i)x^3+(-3/8-6i)x^2+(-929/80+6i)x \
      +3061/160-(15/4)i
R = (-1489/32-(427/10)i)x+17+(3061/40)i
--------------------------------------------------------------
```

7.2 Extended GCD for Polynomials

The extended greatest common divisor algorithm for polynomials $A(x)$ and $B(x)$ asserts the existence of unique polynomials $G(x)$, $S(x)$, and $T(x)$ with G monic such that

- $G(x)$ divides $A(x)$ and $B(x)$.
- If $D(x)$ divides $A(x)$ and $B(x)$ then $D(x)$ divides $G(x)$.
- $G(x) = S(x)A(x) + T(x)B(x)$.

$G(x)$ is called the *greatest common divisor* (gcd) of $A(x)$ and $B(x)$. Note that any scalar multiple of $G(x)$ also has these properties with S and T suitably adjusted. The requirement that G be monic makes G, S, T unique.

The following function takes a pair of polynomials $A(x)$ and $B(x)$ with Gaussian rational coefficients and finds the polynomials $G(x)$, $S(x)$, and $T(x)$. The code is entirely analogous to the version for integers. The difference is in the use of the polynomial version of the division algorithm.

```
def poly_gcd(A,B,var):                      # assumes degree A > degree B
    R0 = A; R1 = B
    S0 = '1'; T0 = '0' # initial values
    while True:
        if i == 0: Q = ''
        G = R0; S = S0; T = T0              # save these: to be returned
        if R1 == '0': break
        Q,R2 = div_alg(R0,R1)               # R0 = Q*R1 + R2
        S2 = S0 +  '-('+ Q +')('+ S1 +')'   # S2 = S0 - Q*S1
        T2 = T0 +  '-('+ Q +')('+ T1 +')'   # T2 = T0 - Q*T1
        S2 = pl.main(S2)[0]
        T2 = pl.main(T2)[0]

        R0 = R1; R1 = R2      # shift
        S0 = S1; S1 = S2      # shift
        T0 = T1; T1 = T2      # shift
        i += 1
    Gmonic,Gmoniclist = pl.main(G)[2:4]     # monic version of G
    Gmultiplier = Gmoniclist[0]
    # divide by Gmultiplier
    S = pl.main('(1/('+ Gmultiplier + '))' + '(' + S + ')')[0]
    T = pl.main('(1/('+ Gmultiplier + '))' + '(' + T + ')')[0]
    if tl.isarithmetic(G):                  # G a constant
        G = '1'
    else:
        G = pl.main('(1/('+ Gmultiplier + '))' + '(' + G + ')')[0]
    return G,S,T
# -------------------------- Sample Run --------------------------
Input:
A = '18x^3 -42x^2 + 30x -6'
B = '-12x^2 + 10x - 2'
G,S,T = poly_gcd(A,B,'x')
print('G = ',G)
print('S = ',S)
print('T = ',T)
# ----------------------------------------------------------------
Output:
G = x-1/3
S = 2/9
T = (1/3)x-1/2
```

7.3 Rational Roots and Linear Factors

A *root* or *zero* of a polynomial $P(x)$ is a value a of x for which $P(a) = 0$. A zero a gives rise to a linear factor $x - a$ of the polynomial. Indeed, by the division algorithm, there exist polynomials $Q(x)$ and $R(x)$ such that

$$P(x) = Q(x)(x - a) + R(x), \quad \deg(R(x)) < \deg(x - a) = 1.$$

If $P(a) = 0$, then R, a constant, must be zero, hence $P(x) = Q(x)(x - a)$.

If $P(x)$ has rational coefficients, we can use the so-called Rational Root Theorem to find the rational zeros of P, if any. The theorem applies to polynomials with integer coefficients, but since any polynomial with rational coefficients is a constant times a polynomial with integer coefficients, there is no loss of generality here. The theorem asserts that if a polynomial

$$P(x) = a_k x^k + a_{k-1} x^{k-1} + \cdots + a_1 x + a_0$$

with integer coefficients has a rational root r, then r must be of the form m/n, where m is a factor of a_0 and n is a factor of a_k. The theorem says nothing about nonrational roots. For example, while the theorem detects the root $x = 1$ of the polynomial $(x - 1)(x^2 - 2)(x^2 + 1)$ it fails to detect the roots $\pm\sqrt{2}, \pm i$. Since there is no general method to find all exact roots of a polynomial, we restrict ourselves to finding rational roots.

The first step is to find the ratios m/n. The following function does this.

```
def form_ratios(N,D):
    # input: positive numerator N, denominator D
    # output: reduced ratios divN/divD as strings
    R = []                                  # list for ratios
    divN = nm.generate_divisors(N)          # list of divisors of N
    divD = nm.generate_divisors(D)          # list of divisors of D
    for i in range(len(divN)):
        for j in range(len(divD)):
            g = ma.gcd(divN[i], divD[j])
            rnum = divN[i]/g        # reduces ratio divN[i]/divD[j]
            rden = divD[j]/g
            r = ar.main(str(rnum) +'/'+ str(rden))[0]
            R.append(r)             # append reduced fraction r
            if r != '0':
                r = tl.fix_signs('-' + r)
                R.append(r)         # append negative of r as well
    R = list(set(R))                # remove duplicates
    return R
```

The function get_roots_and_factors(P) takes a polynomial $P(x)$ and returns its (rational) linear factors, their multiplicity (that is, the power to which a linear factor occurs), the leftover factor that has no rational roots, and the roots themselves.

```
def get_roots_and_factors(pol):
    # output:lists of roots, multiplicity, factors
    var = tl.get_var(pol)
    ipol,ilist = pl.main(pol)[4:6]              # integer coefficients
    L = len(ilist)                              # = degree_pol + 1
    N = ilist[L-1]                              # constant term
    M = ilist[0]                                # multiplier
    D = ilist[1]                                # leading coefficient
    pol = pl.flist2pol(ilist[1:])
    ratios = form_ratios(abs(int(N)),abs(int(D)))   # possible roots
    roots = []
    factors = []
    multiplicity = []
    A = pol                     # dividend for division algorithm
    for r in ratios:
        if A == '0': break
        if ar.evaluate(A,r,'')[0] != '0':       # check if r a root
            continue                            # r not a root; skip iteration
        roots.append(r)                                     # got a root
        if r == '0':                                        # get factor (var-0)
            B = var
        else:
            B = tl.fix_signs(var + '-'+ str(r)) # get factor (var-r)
        m = 0                                   # initialize multiplicity
        while True:                             # keep dividing out factor
            Q, R = div_alg(A,B)
            if R == '0':                        # if B is a factor,
                m += 1                          # then update multiplicity
                A = Q                           # old A with factor B divided out
            else:
                break
        factors.append(B)                               # linear factor B
        multiplicity.append(m)                          # its multiplicity
                        # A now has all linear factor divided out
    A = pl.main( '('+ M +')(' + A +')')[2]              # absorb M into A
    factors = [A] + factors
    multiplicity = [1] + multiplicity # trivial multiplicity for  A
    return roots, multiplicity, factors
```

```
-------------------------- Sample Run --------------------------
Input:
pol = '60x^8-104x^7+7x^6+85x^5-185x^4+227x^3-136x^2+38x-4'
roots, multiplicity, factors = get_roots_and_factors(pol)
print('roots:        ', roots)
print('multiplicity: ', multiplicity)
print('factors:      ',factors)
```

```
Output:
roots:          ['1/2', '2/5', '1/3']
multiplicity:   [1, 1, 1, 2]           # multiplicity of A included
factors:        ['60(x^4-x^2-2)', 'x-1/2', 'x-2/5', 'x-1/3']
```

The function `factor_polynomial(P)` takes a polynomial P and concatenates the factors supplied by `get_roots_and_factors(P)`.

```
def factor_polynomial(pol):
    roots, multiplicity, factors = get_roots_and_factors(pol)
    if len(factors) == 1: return pol
    factorization = ''
    for k in range(len(factors)):    # run through factors
        f = factors[k]
        if f == '1':                                    # trivial factor
            continue
        if '+' in f or '-' in f:
            f = '(' + f + ')'
        m = multiplicity[k]
        if m == '':
            continue
        if m != 1:
            factorization += f + '^' + str(m)
        else:
            factorization += f
    return factorization
```
------------------------------- Sample Run -------------------------------
```
Input:
pol = '60x^8+(-104)x^7+7x^6+85x^5+(-185)x^4+227x^3+(-136)x^2+38x-4'
factorization = factor_polynomial(pol)
#check:
pl.main(factorization)[0]
```

```
Output:
60(x^4-x^2-2)(x-2/5)(x-1/2)^2(x-1/3)
60x^8-104x^7+7x^6+85x^5-185x^4+227x^3-136x^2+38x-4
```

7.4 Modular Division Algorithm

The division algorithm for polynomials with integer coefficients mod p is similar to that for polynomials with rational coefficients. The difference is in the operations that require division by coefficients. This is reflected in the calculation of the quantity `reciprocal` in the code. The calculation uses the inversion function `mod_mult_inv` from Chap. 4, which returns -1 if inversion is not possible.

```
def div_alg_mod(A,B,p): # A,B polynomial strings, p positive integer
    #returns quotient and remainder
    var = tl.get_var(A)
    AL = pl.main_mod(A,p)[1]           # reduced polynomial list of A
    BL = pl.main_mod(B,p)[1]           # reduced polynomial list of B
    if BL[0] == '0': return
    degree_A = len(AL) - 1
    degree_B = len(BL) - 1
    if degree_A < degree_B:                      # trivial case
        Q = '0';
        R = A;
        return Q, R
    if degree_B == 0:
        reciprocal = str(nm.mod_mult_inv(int(BL[0]),p))
        if reciprocal == '-1':
            return 'none','none'           # division not possible
        Q = pl.poly_mod('('+ A +')('+ reciprocal +')',p)[0]
        R = '0'
        return Q, R
    Q = ''    # initialize quotient string
    while True:
        degree = len(AL) - len(BL)
        reciprocal = str(nm.mod_mult_inv(int(BL[0]),p))
        if reciprocal == '-1':
            return 'none','none'           # division not possible
        q = '('+ AL[0] +')('+ reciprocal +')'
        q = pl.strmod(ar.main(q)[0],p)             # q = AL[0]/BL[0]
        # attach power to q:
        Q = Q + '+(' + '('+q+')' + var + '^' + str(degree) + ')'
        Q,QL = pl.poly_mod(Q,p)[0:2]
        CL = pl.pol_scalar_prod(q,BL)              # C = q*B
        # attach 0's to DL to match degree of AL
        CL = CL + ['0' for k in range(len(AL) - len(BL))]
        AL = pl.pol_diff(AL,CL)
        AL = pl.integerlist2modlist(AL,p)             # reduce
        if len(AL)< len(BL): break                    # done
    RL = AL
    R = pl.flist2pol(RL)           # convert list to polynomial
    return Q, R
```

--------------------------- Sample Run ---------------------------
Input:
A = 'x^5-2x^3+7x^2-4x+11'
B = '4x^2+3x+4'
output_list = []
for p in range(2,12):   # calculate mod p Q,R for p = 2 to 11
    Q,R = div_alg_mod(A,B,p)
    row = ['mod',str(p),' Q =',Q,' R =',R]
    output_list.append(row)
tl.format_print(output_list, 3, 'left')

```
Output:
mod  2    Q =  x^4+x                    R =  1
mod  3    Q =  x^3+1                    R =  2x+1
mod  4    Q =  3x^4+2x^2+x              R =  3
mod  5    Q =  4x^3+2x^2+4x+3           R =  x+4
mod  6    Q =  none                     R =  none
mod  7    Q =  2x^3+2x^2+3x+1           R =  2x
mod  8    Q =  none                     R =  none
mod  9    Q =  7x^3+6x^2+6x+7           R =  5x+1
mod  10   Q =  none                     R =  none
mod  11   Q =  3x^3+6x^2+3x+10          R =  9x+4
```

7.5 Modular Extended Greatest Common Divisor

The function `poly_gcd_mod(A,B,p)` is the modular analog of `poly_gcd`. It takes polynomials A, B with integer coefficients and returns the gcd $G(x)$ of $A(x)$ and $B(x)$ as well as polynomials $S(x)$ and $T(x)$, all with coefficients modulo p, such that $G(x) = S(x)A(x) + T(x)B(x)$.

```
def poly_gcd_mod(A,B,p):              # assumes degree A > degree B
    R0 = A; R1 = B
    S0 = '1'; T0 = '0' # initial values
    S1 = '0'; T1 = '1'
    i = 0
    while True:
        G = R0; S = S0; T = T0         # save: returned by function
        if R1 == '0': break
        Q,R2 = div_alg_mod(R0,R1,p)              # R0 = Q*R1 + R2
        if Q == 'none' or R2 == 'none':
            return 'none','none','none'
        S2  = S0 +   '-('+ Q +')('+ S1 +')'      # S2 = S0 - Q*S1
        T2  = T0 +   '-('+ Q +')('+ T1 +')'      # T2 = T0 - Q*T1
        S2 =pl.poly_mod(S2,p)[0]
        T2 =pl.poly_mod(T2,p)[0]
        R0 = R1; R1 = R2                                     # shift
        S0 = S1; S1 = S2                                     # shift
        T0 = T1; T1 = T2                                     # shift
        i += 1
------------------------- Sample Run -------------------------
Input:
A = '8x^6+14x^5+12x^4+61x^3+70x^2+60x+105'
B = '10x^5+18x^4+35x^3+63x^2+30x+54'
output_list = []
for p in range(2,12):
    G,S,T = poly_gcd_mod(A,B,p)
```

```
        row = ['mod',str(p),'G =',G,'S =',S,'T =',T]
        output_list.append(row)
    tl.format_print(output_list, 2, 'left')
```
```
Output:
mod  2   G =  x+1                S =  x+1         T =  x
mod  3   G =  2x^4+x^2            S =  1           T =  x+1
mod  4   G =  none               S =  none        T =  none
mod  5   G =  4x^2+1              S =  x+4         T =  4x^3+3x^2+4
mod  6   G =  none               S =  none        T =  none
mod  7   G =  2x^2+3              S =  5x^2+6x+1   T =  3x^3+x^2+4x+2
mod  8   G =  none               S =  none        T =  none
mod  9   G =  none               S =  none        T =  none
mod  10  G =  none               S =  none        T =  none
mod  11  G =  6x^3+7x^2+9x+5      S =  x+3         T =  8x^2+6x+2
```

7.6 Modular Roots and Factors

To find the roots of a polynomial $P(x)$ with integer coefficients mod p one need only run through the remainders $r = 0, 1, 2, \ldots p - 1$, checking if $P(r) = 0$.

```
def get_mod_roots(pol,var,p):
    roots = []
    for r in range(1,p):
        if pl.pol_mod_eval(pol,str(r),p) == 0:
            roots.append(r)                        # got a root
    return roots
```

The function get_mod_factors(P,p) is the analog of get_factors(P).

```
def get_mod_factors(P,p):
# output: roots, multiplicity, factors lists
    var = tl.get_var(P)
    lin_factors = []
    multiplicity = []
    A = pl.poly_mod(P,var,p)[0]
    roots = get_mod_roots(A,var,p)                 # get all roots first
    if roots == []: return P
    for r in roots:
        if r == 0:
            B = var
        else:
            B = tl.fix_signs(var + '-'+ str(r))
        m = 0                                      # initialize multiplicity
```

7.6 Modular Roots and Factors

```
        while True:                         # keep dividing out factor B
            Q, R = div_alg_mod(A,B,p)
            if R == '0':                    # if B a factor,
                m += 1                      # update its multiplicity
                A = Q                       # old A with factor B divided out
            else: break

        # A now has all linear factors divided out
        lin_factors.append(B)               # append linear factor
        multiplicity.append(m)              # and its multiplicity
    leftover_factor = A
    return lin_factors, multiplicity, leftover_factor
```

The function `factor_pol_mod(P,p)` is the analog of the function `factor_pol(P)`. It takes a polynomial *P* and an integer *p* > 1 and returns its mod *p* factorization.

```
def factor_pol_mod(P,p):
    lin_factors, multiplicity, leftover = get_mod_factors(P,p)
    factorization = ''
    for k in range(len(lin_factors)):
        lf = lin_factors[k]
        if '+' in lf or '-' in lf:
            lf = '(' + lf + ')'
        m = multiplicity[k]
        if m == 0:
            continue
        if m != 1:
            factorization += lf + '^' + str(m)
        else:
            factorization += lf
    if not tl.isarithmetic(leftover):
        leftover = '(' + leftover + ')'
    if leftover == '1':
        leftover = ''
    return leftover + factorization
```

---------------------------- Sample Run ----------------------------
```
Input:
P = '48x^6+(-356)x^5+984x^4+(-1361)x^3+1286x^2+(-1005)x+350'
for p in range(2,13):
    factorization = factor_pol_mod(P,p)
    print('mod',p,': ',factorization)
```

```
Output:
mod 2:   x(x-1)^2
mod 3:   (x^2+1)(x-1)^3
mod 4:   (3x^2+3)(x-2)
```

```
mod  5:  3x^2(x-2)(x-3)^2(x-4)
mod  6:  (4x^2+1)(x-1)^2(x-4)
mod  7:  (6x^2+6)x(x-3)(x-6)^2
mod  8:  (4x^4+7x^2+3)(x-6)
mod  9:  (3x^3+4x^2+3x+4)(x-1)^2(x-7)
mod 10:  (8x^3+2x^2+4x+5)x(x-3)^2
mod 11:  (4x^2+4)(x-8)^3(x-10)
mod 12:  (4x^4+4x^3+11x^2+4x+7)(x-10)
```

Multivariable Algebra

8

In this chapter we construct the module `MultiAlg.py`, which takes an expression involving monomials of several variables with Gaussian rational coefficients (also called scalars) and converts it into a single rational function. The module is headed by the import statements

```
------------------------- MultiAlg.py -------------------------
import Arithmetic as ar
import Tools as tl
import Number as nm
from operator import itemgetter
---------------------------------------------------------------
```

8.1 Rational Functions and Their Representations

A *monomial in several variables*, or simply *monomial*, is an expression consisting of letters (*variables*) raised to positive integer powers and a complex number called a *coefficient*. Powers and coefficients equal to 1 are omitted as usual. The *degree* of a monomial is the sum of its exponents. For example, $2y^3 x^2 z$ is a monomial in the variables x, y, z with coefficient 2 and degree 6. Monomials are multiplied by taking the products of the coefficients and adding exponents of like variables. For example,

$$(3 x^3 y^2 z)(2.1 \, i \, xyz) = 6.2 \, i \, x^4 y^3 z^2.$$

We consider constants c as monomials, as in $c = cx^0 y^0 z^0$.

© The Author(s), under exclusive license to Springer Nature Switzerland AG 2025
H. D. Junghenn, *Symbolic Mathematics with Python*, Synthesis Lectures on Mathematics & Statistics, https://doi.org/10.1007/978-3-031-90522-3_8

A *polynomial in several variables* or *multivariate polynomial* is a sum of monomials of several variables, called the *terms* of the polynomial. The *degree* of a polynomial is the largest degree of its terms. Multivariate polynomials, like polynomials of a single variable, are added, subtracted, multiplied, and raised to powers by the usual algebraic techniques: multiplying the monomials and collecting like terms, for example,

$$3x^2(zy + xy) + 5yx^3 = 3x^2yz + 8x^3y.$$

We order the terms of a polynomial by the degree of the monomials, starting with the highest degree. The variables within monomials are ordered alphabetically for readability. The right side in the preceding example adheres to these conventions.

A *rational function* of several variables is a ratio of multivariate polynomials. Rational functions may be added, subtracted, multiplied or divided like ordinary fractions. For example, if P, Q, R, S are polynomials, then

$$\frac{P}{Q}\frac{R}{S} = \frac{PR}{QS}$$

and

$$\frac{P}{Q} + \frac{R}{S} = \frac{PS + RQ}{QS}.$$

Note that a polynomial P may be viewed as the rational function $P/1$. This will be useful in calculations, as it allows polynomials and rational functions to be treated on an equal basis.

A monomial is represented in the module by a list containing the coefficient and exponents of the variables, the list obtained during run time. For example, if the variable list is `['x','y','z']`, then the monomial $(3.1 + 2.5i)z^6x$ is represented by the list `['3.1+2.5i',1,0,6]`. A polynomial is represented as a list of monomial lists. For example, for the preceding variable list, the monomial $2z^3xy^2 + 3y$ is represented by the double list

`[['2', 1, 2, 3], ['3', 0, 1, 0]].`

A rational function is represented by two double lists, the first of which is numerator list and the second the denominator list. For example, for the above variable list, the expression $(2z^3xy^2 + 3y)/(7z^6x^4 - 13)$ is represented by the list

`[ [['2', 1, 2, 3], ['3', 0, 1, 0]], [['7', 4, 0, 6],['-13', 0, 0]] ].`

While such nested lists may seem overly intricate, they turn out to be well-suited for carrying out complicated algebraic operations involving rational functions, multilayered as these operations are.

8.2 Overview

Here is the code for the main function of the module. The output is a rational expression with Gaussian rational coefficients, as well as one with Gaussian integer coefficients and a compensating multiplier. Also returned are the lists for these expressions.

```
def main(expr):
    global idx                      # points to current character in expr
    global varlist                  # list of variable names in expr
    varlist = tl.get_vars(expr)[0]  # extract variable list
    if varlist == []:               # no varlist?
        z,c = ar.main(expr)         # then use arithmetic
        return z,c,'',''
    expr = tl.attach_missing_exp(expr,varlist)
    expr = tl.fix_signs(expr)
    expr = tl.fix_operands(expr)
    expr = tl.insert_asterisks(expr,varlist)
    idx = 0                         # point to beginning of expr
    R = allocate_ops(expr,0)        # do the calculations
    num = R[0]; den = R[1]
    num = combine_monos(num)        # simplify polynomial
    den = combine_monos(den)
    num = sort_list(num)
    den = sort_list(den)
    ratlist = [num,den]
    rat = list2rational(ratlist)
    irat, iratlist = list2int_rational(ratlist)
    return rat,ratlist,irat,iratlist
```

```
--------------------------- Sample Run ---------------------------
Input:
expr = 'z_2/(x-1.1) + 3x/(y1-3.2i)'
rat,rat_list, irat, irat_list = main(expr)
print(rat)
print(irat)
------------------------------------------------------------------
Output:
(3x^2+y1z_2+(-33/10)x+(-16/5)iz_2)/(xy1+(-11/10)y1+(-16/5)ix+(88/25)i)
5(30x^2+10y1z_2-33x-32iz_2)/(50xy1-55y1-160ix+176i)
------------------------------------------------------------------
```

Notice that the function allows subscripted variables, in this case y1. The only condition is that one cannot have both subscripted and unscripted variables with the same letter. For example, the variables x and x1 are not allowed in the same expression, as this confuses the function get_vars.

8.3 Combining Monomials

Calculations are performed on the multilists described earlier. The following function is central to these calculations. It combines monomials with like powers into a polynomial list.

```
def combine_monos(P):
    # eg. [['2',3,4],['5',3,4],['-3',8,9]]-->[['7',3,4],['-3',8,9]]
    Q = []                               # for reduced polynomial
    if len(P) == 1:
        return P
    for i in range(len(P)-1):
        if P[i] == '': continue          # already added
        M = P[i]                         # ith monomial: [coeff,powers]
        # add succeeding like monomials to M:
        for j in range(i+1,len(P)):
            if P[j] == '':
                continue
            N = P[j]
            if M[1:] == N[1:]:   # if powers same, add coefficients
                coeffsum = ar.main(M[0] + '+(' + N[0] +')')[0]
                M[0] = coeffsum          # update M's coefficient
                P[j] = ''                # mark as already added
        Q.append(M)                 # append nonzero monomial in P[i]
    leftover_mono = P[len(P)-1]
    if leftover_mono != '' and leftover_mono[0] != '0':
        Q.append(leftover_mono)    # pick up leftover monomial at end
    return Q
```

To see how the function works, consider the list

$$P = [['2',3,4], ['4',8,9], ['5',3,4],['-3',8,9], ['11',3,4]],$$

which, for the variable list ['x','y'], represents the polynomial

$$2x^3y^4 + 4x^8y^9 + 5x^3y^4 - 3x^8y^9 + 11x^3y^4.$$

The function takes the first monomial's list, ['2',3,4], adds to it all monomial lists with the same exponents, and then marks the latter with a null string to indicate that it has already been added. This process results in the new list

$$P = [['18',3,4], ['4',8,9], '',['-3',8,9], ''].$$

The latter list then undergoes the same process, resulting in the final list

$$P = [['18',3,4], ['1',8,9],'' ,'',''].$$

The function returns [['18',3,4], ['1',8,9]].

8.4 Scalar and Variable Conversion Functions

The following functions are used to convert scalars to rational multilists.

```
------------------------------------------------------------------
def scalar2mono(s):                    # attach scalar to zero list
    return [s] + [0 for k in range(len(varlist))]

def scalar2rat(s):
    return [[scalar2mono(s)], [scalar2mono('1')]]

-------------------------- Sample Run ----------------------------
Input:
varlist = list('xyz')
print(scalar2mono('2.3/i'))
------------------------------------------------------------------
Output:
[[['2.3/i', 0, 0, 0]], [['1', 0, 0, 0]]]
------------------------------------------------------------------
```

The next functions are used to convert a variable to a rational multilist. They are analogous to the above scalar conversion functions.

```
------------------------------------------------------------------
def var2mono(var,exp):                 # variable and its exponent
    # e.g. 'x^2' -> ['1',2,0,...]
    M = scalar2mono('1',len(varlist))
    position = varlist.find(var)
    M[position+1] = exp        # put exponent in correct position
    return M

def var2rat(var,exp):
    num = var2mono(ch,exp)
    den = scalar2mono('1',len(varlist))
    R = [[num], [den]]

-------------------------- Sample Run ----------------------------
Input:
varlist = list('xyz')
print(var2rat('y',7))
------------------------------------------------------------------
Output:
[[['1', 0, 7, 0]], [['1', 0, 0, 0]]]                    # = y^7
------------------------------------------------------------------
```

8.5 Calculations

The functions in this section do the actual calculations, following the natural hierarchy of operations in a rational function. The first function takes two monomial lists, multiplies their coefficients, and adds the exponents.

```
def multiply_monos(M,N):          # returns product of monomials M,N
    K = []                        # output list
    # multiply coefficients M[0], N[0]:
    coeff_prod = ar.main('('+ M[0] +')('+ N[0] +')')[0]
    K.append(coeff_prod)
    for i in range(1,len(N)):     # add integer exponents of like vars
        K.append(M[i] + N[i])
    return K

-------------------------- Sample Run --------------------------
Input:
M = ['2',3,4,5]                                    # 2x^3y^4z^5
N = ['-3',6,7,8]                                   # -3x^6y^7z^8
print(multiply_monos(M,N))

Output:
['-6', 9, 11, 13]                                  # -6x^9y^11z^13
```

The function `multiply_pols(P,Q)` takes two polynomial multilists P,Q, multiplies all possible pairs M,N of the monomial sublists, with M in P and N in Q, and then simplifies the result by combining monomials.

```
def multiply_pols(P,Q):           # multiplies 2 lists of monomials
    R = []                        # list for product polynomial
    for M in P:                   # get all possible products M*N
        for N in Q:
            K = multiply_monos(M,N)
            R.append(K)           # append product to pol list R
    R = combine_monos(R)          # collect terms
    return R

-------------------------- Sample Run --------------------------
Input:
P = [['2',3,4,5], ['-1',6,7,8]]
Q = [['5',9,10,11], ['7',12,14,16]]
print(multiply_pols(P,Q))

Output:
[['10', 12, 14, 16], ['14', 15, 18, 21], ['-5', 15, 17, 19], \
 ['-7', 18, 21, 24]]
```

8.5 Calculations 137

The function multiply_rationals(R,S) multiplies the numerators R[0],S[0] and denominators R[1],S[1] of the rational lists R,S, thus forming the numerator and denominator of the product. The function divide_rationals(R,S) inverts S and multiplies the result by R.

```
def multiply_rationals(R, S):      # R = [R[0],R[1]] S = [S[0],S[1]]
    num = multiply_pols(R[0], S[0])         # multiply numerators
    den = multiply_pols(R[1], S[1])         # multiply denominators
    return [num,den]

def divide_rationals(R,S):
    T = [S[1],S[0]]                          # invert S
    return multiply_rationals(R,T)
```

The function rational_power(R,n) takes a rational list R and an integer n and returns the power R^n.

```
def rational_power(R,n):
    need_to_invert = False
    if n == 1: return R                      # no exponentiation
     if n == 0:
        num = ['1'] + [0 for k in range(len(R[0][0])-1)]
        den = num
        return [num,den]    # returns rational list representing '1'
    if n < 0:
        n = -n
        need_to_invert = True
    S = R
    for i in range(n-1):       # multiply R times itself n-1 times
        S = multiply_rationals(S, R)
    num = S[0]); den = S[1]
    if need_to_invert:
        return [den,num]
    return [num,den]
```

The following group of functions culminates in one that multiplies a rational list by a scalar. For example, if s ='1/2' and

R = [[['2',3,4,5], ['-1',6,7,8]], [['5',9,10,11], ['7',12,13,14]]], then the function produces the list

```
[[['1',3,4,5], ['-1/2',6,7,8]], [['5',9,10,11], ['7',12,13,14]]].
```

```
def mono_scalar_prod(s,M):
    # e.g. '3'*['5',6,7] = ['15',6,7]
    prod = ar.main('(' + s +')(' + M[0] + ')')[0]    # s times coeff
    return [prod] + M[1:]

def pol_scalar_prod(s,P):
    # e.g. '3'*[['5',6,7],['7',8,9]] = [['15',6,7], ['21',8,9]]
    Q = []
    for M in P:
        N = mono_scalar_prod(s,M)            # multiply each mono by s
        Q.append(N)
    return Q

def rat_scalar_prod(s,R):
    return [pol_scalar_prod(s,R[0]),R[1]]    # multiply numerator by s
```

The last three functions in this section add and subtract rationals.

```
def add_pols(P,Q):
    if P == []: return Q
    if Q == []: return P
    S = P+Q                                  # concatenate lists
return combine_monos(S)                      # tidy up

def add_rationals(R,S):          # R = [R[0],R[1]], S = [S[0],S[1]]
    numR = R[0]; denR = R[1]; numS = S[0]; denS = S[1]
    A = multiply_pols(denR,numS)
    B = multiply_pols(denS,numR)
    num = add_pols(A,B)                      # denR*numS + denS*numR
    num = combine_monos(num)
    den = multiply_pols(denR,denS)           # denR*denS
    den = combine_monos(den)
    return [num,den]             # (denR*numS + denS*numR)/denR*denS

def subtract_rationals(R, S):
    minusS = rat_scalar_prod('-1',S)
    return add_rationals(R,minusS)
```

8.6 The Allocator

The function `allocate_ops(expr,mode)` assigns the computational tasks to the appropriate function of the preceding section. It is similar to the eponymous function in Chap. 6.

```
def allocate_ops(expr,mode):
    global idx, varlist
    R = []
    while idx < len(expr):
        ch = expr[idx]
        if tl.isnumeric(ch):
            start = idx
            s,idx = tl.extract_numeric(expr, start)
            R =  scalar2rat(s)
        elif tl.isletter(ch):
            var,idx = tl.extract_var(expr,idx)
            R = var2rat(var,1)
        elif ch == '+':
            if mode > 0: break          # wait for higher mode
            idx += 1
            S = allocate_ops(expr,0)
            R = add_rationals(R,S)
        elif ch == '-':
            if mode > 0: break          # wait for higher mode
            idx += 1
            S = allocate_ops(expr,1)
            R = subtract_rationals(R,S)
        elif ch == '*':
            if mode > 1: break
            idx += 1
            S = allocate_ops(expr,1)
            R = multiply_rationals(R,S)
        elif ch == '/':
            if mode > 1: break
            idx += 1
            S = allocate_ops(expr,1)
            R = divide_rationals(R,S)
        elif ch == '^':
            exp,idx = tl.extract_exp(expr,idx)
            exp = ar.main(exp)[0]
            R = rational_power(R, int(exp))
        elif ch == '(':
            start = idx
            paren_expr,end = tl.extract_paren(expr,start)
            if  tl.isarithmetic(paren_expr):
                r = ar.main(paren_expr)[0]
                num = scalar2mono(r)
                den = scalar2mono('1')
                R = [[num],[den]]
                idx=end
            else:
```

```
                idx+=1
                R = allocate_ops(expr,0)
                idx+=1
        elif ch == ')': break
    return R
```

8.7 Sorting

The function sort_list(P) sorts the terms of a polynomial list P by degree, which, recall, is the largest degree of its monomials. For each monomial list, the function adds the exponents and then attaches the sum to the beginning of the list. The polynomial list is then sorted on this field. For example, if the polynomial list is

$$[['2', 1, 0, 3], ['5', 3, 1, 6], ['7', 0, 3, 2]],$$

then the function first generates the list

$$[[4,'2', 1, 0, 3], [10,'5', 3, 1, 6], [5,'7', 0, 3, 2]].$$

The first item of each sublist is the sum of the exponents. Next, the function sorts on these sums, producing

$$[[10,'5', 3, 1, 6], [5,'7', 0, 3, 2], [4,'2', 1, 0, 3]].$$

Finally, the function removes the sums in the first entry:

$$[['5', 3, 1, 6], ['7', 0, 3, 2], ['2', 1, 0, 3]].$$

The sorting is done by function itemgetter from the Python package operator, which allows sorting on any field, in our case the first one.

```
def sort_list(P):                       # polynomial list
    l = len(P)                          # number of mono lists in pol
    m = len(P[0])                       # length of a mono list
    if l == 1:
        return P
```

```
        Q = tl.copylist(P)                        # don't change P
        for i in range(l):             # sum the powers of each mono list
            sum = 0
            for j in range(1,m):            # sum the powers of ith mono
                sum = sum + Q[i][j]
            Q[i] = [sum] + Q[i]         # attach sum to start of ith mono
        # sort pol_list on the first field (sum of powers):
        sorted_list = sorted(Q, key=itemgetter(0), reverse = True)
        for i in range(l):         # remove sum of powers from each mono
            # extract 1st part of the mono sorted_list[i]:
            sorted_list[i] = sorted_list[i][1:]
        return sorted_list
```

8.8 Coefficient Reduction

The function reduce_coeffs(R) takes a rational list R with integer coefficients and cancels common factors in the coefficients of the numerator and denominator.

```
def reduce_coeffs(R):
    num = R[0]; den = R[1]
    numden = num + den                              # combine lists
    coeffs = []
    for i in range(len(numden)):
        n = numden[i][0]           # get the coefficient in ith list
        re = int(ar.real(n))
        im = int(ar.imag(n))
        coeffs.append(re)          # append real and imaginary parts
        coeffs.append(im)                          # to coeff_list
    gcd = nm.multi_extended_gcd(coeffs)[0]   # largest common factor
    for i in range(len(num)):      # divide numerator coeffs by gcd
        n = num[i][0]
        num[i][0] = ar.main('('+ n +')/('+ str(gcd) +')')[0]
    for i in range(len(den)):      # divide denominator coeffs by gcd
        n = den[i][0]
        den[i][0] = ar.main('('+ n +')/('+ str(gcd) +')')[0]
    return [num,den]

-------------------------- Sample Run --------------------------
Input:
R = [ [['2+10i',3,4,5],['4',3,7,8]], [['6',3,4,9],['8',3,8,11]] ]
print(R)
----------------------------------------------------------------
Output:
[ [['1+5i',3,4,5], ['2',3,7,8]], [['3',3,4,9], ['4',3,8,11]] ]
```

In the sample run and for `varlist = ['x','y','z']`, the rational function

$$\frac{(2+10i)x^3y^4z^5 + 4x^3y^7z^8}{6x^3y^4z^9 + 8x^3y^8z^{11}}$$

is reduced to

$$\frac{(1+5i)x^3y^4z^5 + 2x^3y^7z^8}{3x^3y^4z^9 + 4x^3y^8z^{11}}$$

8.9 Variable Reduction

The function `reduce_vars(R)` takes a rational list and cancels common variables in the numerator and denominator lists. It invokes, for each k, the function `get_smallest_exp(R,k)`, which returns the minimum exponent in position k of all the monomials comprising the list R, and then subtracts that exponent from the monomial exponents in position k. Here's the code:

```
def get_smallest_exp(R,k):
    # returns minimum of all exponents in R in kth monomial position
    num = R[0]; den = R[1]
    min_exp = 100000                        # arbitrary but start high
    for M in num:
        if min_exp > M[k]: min_exp = M[k]
    for M in den:
        if min_exp > M[k]: min_exp = M[k]
    return min_exp

def reduce_vars(R):
    num = R[0]; den = R[1]
    L = len(num[0])                                    # monomial length
    for k in range(1,L):   # k = position of kth variable's exponent
        min_exp = get_smallest_exp(R,k)
        for M in num:            # run through monomials in numerator
            M[k] = M[k] - min_exp       # reduce kth exponent in num
        for N in den:            # run through monomials in denominator
            N[k] = N[k] - min_exp       # reduce kth exponent in den
    return [num,den]

---------------------------- Sample Run ---------------------------
Input:
R = [ [['2',2,5],['5',7,8]], [['7',3,4],['9',2,11]] ]
print(reduce_vars(R))
-------------------------------------------------------------------
Output:
[ [['2', 0, 1], ['5', 5, 4]], [['7', 1, 0], ['9', 0, 7]] ]
-------------------------------------------------------------------
```

In the sample run and for `varlist = ['x','y']`, the rational function

$$\frac{2x^2y^5 + 5x^7y^8}{7x^3y^4 + 9x^2y^{11}}$$

is reduced to

$$\frac{2y + 5x^5y^4}{7x + 9y^7}$$

8.10 Clearing Fractions

The function `get_pol_lcm(P)` returns the lcm of all the coefficient denominators in the polynomial *P*. For this it uses `mono_denoms(M)` to get the denominators of the real and imaginary parts of `M[0]`.

```
-------------------------------------------------------------------
def get_pol_lcm(P):   # get lcm of all denominators of monomials in P
    pol_denoms = []
    for M in P:
        mono_denoms = get_mono_denoms(M)   # denominators M's coeff
        pol_denoms = pol_denoms + mono_denoms
    return nm.listlcm(pol_denoms)

def get_mono_denoms(M):
    mono_denoms = [1]          # default: ensures nonempty list
    coeff = M[0]
    re = ar.real(coeff)        # get real and imaginary parts of coeff
    im = ar.imag(coeff)
    if '/' in re:              # get real part denominator
        mono_denoms += [int(re.split('/')[1])]
    if '/' in im:              # imaginary part denominator
        mono_denoms += [int(im.split('/')[1])]
    return mono_denoms
-------------------------------------------------------------------
```

The function `clear_pol(P)` takes a polynomial list and multiplies the coefficients by the least common multiple of their denominators.

```
-------------------------------------------------------------------
def clear_pol(P):
    Q = []
    lcm = str(get_pol_lcm(P))   # lcm of all coeff denominators in P
    for M in P:                 # multiply each coefficient M[0] by lcm
        prod = ar.main( lcm +'('+ M[0] +')' )[0]
        N = [prod] + M[1:]      # concatenate product and exponents of M
        Q.append(N)
    return lcm, Q
------------------------- Sample Run -----------------------------
```

```
Input:
P = [['3/2',3,4,5], ['5/12i',3,7,8], ['7/3+(5/6)i',3,4,9]]
lcm, Q = clear_pol(P)
print(lcm)
print(Q)
```

```
Output:
12
[['18', 3, 4, 5], ['-5i', 3, 7, 8], ['28+10i', 3, 4, 9]]
```

In the sample run, the lcm of the denominators 2, 12, 3, 6 is 12. The output list is that of M but with the coefficient of M multiplied by the lcm. This clears the fractions but of course changes M. The lcm is returned to commemorate the operation, allowing the recovery of M.

8.11 Converting a List into a Rational Function

The final task of the module is the conversion of the rational list returned by allocate_ops into a formatted rational function. This requires several steps. The first is carried out by the function list2monomial(M), which converts a monomial list M into the monomial it represents. For example, if the list of variables is ['x','y','z'], then ['2i',3,4,5] is converted into the monomial '2ix^3y^4z^5'.

```
def list2monomial(Mlist):
    # takes monomial list Mlist and returns the monomial
    coeff = Mlist[0]
    if coeff == '0' or coeff == '-0':
        return '0'
    mono = ''                              # initialize mono string
    for k in range(1,len(Mlist)):
        var = varlist[k-1]                 # run through the variables
        exp = Mlist[k]                     # exponent of variable
        if exp == 0: continue              # omit variable with zero exponent
        mono = mono + var                  # attach variable
        if exp != 1:                       # and exponent if != 1
            mono = mono + '^' + str(exp)
    if coeff == '1' and mono != '':
        coeff = ''        # coeff '1' superfluous for nontrivial mono
    if coeff == '-1' and mono != '':
        coeff = '-'
    if coeff != '' or exp != 0:            # if monomial is not a constant
        coeff = tl.add_parens(coeff)       # add suitable parens
    return tl.fix_signs(coeff + mono)      # attach coeff to mono
```

8.11 Converting a List into a Rational Function

The function list2polnomial(P) converts a polynomial list P, that is, a list of monomial lists, into the required polynomial. For example, if the list of variables is ['x','y'], then the polynomial list P = [['-2',3,4],['5',6,7]] is converted into the polynomial '5x^6y^7-2x^3y^4'.

```
def list2polynomial(P):
    Q = sort_list(P)
    pol = ''
    for M in Q:
        mono = list2monomial(M)           # get the monomial
        if mono == '0': continue          # skip any zero coeff
        pol = pol + '+' + mono            # concatenate to pol
    if pol == '': pol= '0'
    pol = tl.fix_signs(pol)
    return pol
```

The function list2rational(R) converts the numerator and denominator of the rational list R into formatted polynomials and then returns the ratio.

```
def list2rational(R):
    R = reduce_vars(R)                    # simplify
    num = list2polynomial(R[0])           # convert numerator list to pol
    den = list2polynomial(R[1])           # convert denominator list to pol
    if num == '0': return '0'
    if den == '1' or den == '': return num
    if den == '-1':
        num = tl.add_parens(num)
        return '-' + num
    num = tl.add_parens(num)
    den = tl.add_parens(den)
    return num + '/' + den                # return ratio of pols
```

------------------------- Sample Run -------------------------
Input:
R = [[['2',3,4,5],['-1',6,7,8]], [['5+i',9,10,11],['7',12,14,16]]]
varlist = 'xyz'
print(list2rational(R))

Output:
(-x^3y^3z^3+2)/(7x^9y^10z^11+(5+i)x^6y^6z^6)

8.12 Rational Function with Integer Coefficients

The function list2int_rat(R) takes a rational list R and returns an equal rational function with integer coefficients. It does so by multiplying the coefficients of the monomials in the numerator R[0] by the least common multiple of the coefficient denominators, doing the same thing for denominator R[1], and then attaching the compensating multiplier.

```
def list2int_rational(R):
    # converts rational list into integer rational function
    numR = R[0];  denR = R[1]
    if numR == denR: return '1',''
    lcmP,P = clear_pol(numR)
    lcmQ,Q = clear_pol(denR)
    factor = ar.main(lcmQ +'/'+ lcmP)[0]
    irat_list = [P,Q]
    irat_list = reduce_coeffs(irat_list)
    irat = list2rational(irat_list)
    if factor == '1': factor = ''
    factor = tl.add_parens(factor)
    irat = tl.add_parens(irat)
    return factor + irat,irat_list
```

------------------------- Sample Run ------------------------
```
Input:
R = [ [['1/2',2,3], ['1/3',4,5]], [['1/4+i/5',6,7], ['1/6',8,9]] ]
varlist = ['x','y']
print(list2int_rational(R)[0])
```
```
Output:
10(3+2x^2y^2)/((15+12i)x^4y^4+10x^6y^6)
```

8.13 Evaluating an Expression

The function evaluate(expr,substitutions,p) takes a rational expression, a string of substitutions for some or all of its variables, and a decimal length p and returns the corresponding decimal value of the expression.

```
def evaluate(expr,substitutions,p):
    if substitutions == '':
        return expr
    substitutions = substitutions.split(',')
    for i in range(len(substitutions)):
        substitutions[i] = substitutions[i].replace(' ','')
        var,val = substitutions[i].split('=')
        if var == '' or var not in expr: continue
```

8.14 Application: Partial Differentiation of Rational Functions

```
            expr = expr.replace(var,'(' + val + ')')
        expr = tl.fix_signs(expr)

        # decimal approximation if expr is an arithmetic expression:
        if tl.isarithmetic(expr) and p != '':
            return ar.decimal_approx(expr,p)[0]
        return main(expr)[0]

    -------------------------- Sample Run --------------------------
    Input:
    e = '2 + xy^3 + yz + 3z^2 + x/(2z+y)'
    print(evaluate(e,'x=3,y=4,z=5',5))           # decimal value
    print(evaluate(e,'x=3,y=4,z=5',''))          # fractional value
    print(evaluate(e,'x=3,y=4',''))              # polynomial in z
    print(evaluate(e,'x=3',''))                  # polynomial in y,z
    ----------------------------------------------------------------
    Output:
    289.21428
    4049/14
    (6z^3+20z^2+404z+779)/(2z+4)
    (3y^4+6y^3z+5yz^2+6z^3+y^2z+2y+4z+3)/(y+2z)
    ----------------------------------------------------------------
```

8.14 Application: Partial Differentiation of Rational Functions

The main function in this section finds partial derivatives of arbitrary orders of rational functions of several variables. It does so by finding the derivatives of the monomials comprising the function and then assembling these according to the rules for differentiating a rational function.

Monomial Differentiation

The following function finds the derivative of a monomial with respect to a specified variables. The function takes the list representing the monomial and the position of the variable's exponent in the list, multiplies the constant term by the exponent in the list and reduces that exponent by one. For example, if the monomial list is ['1.2',3,4,5] (representing the monomial $1.2\,x^3 y^4 z^5$ for the variable list ['x','y','z']) and the position in the list is 2, (exponent = 4), then the function returns the monomial list ['4.8',3,3,5], having multiplied the coefficient 1.2 by the exponent 4 and reduced the exponent to 3.

```
def der_mono(Mlist,var_position):
    L = len(Mlist)
    # if variable is missing or monomial a constant, return zero
    if Mlist[var_position] == 0 or is_constant(Mlist):
        return ['0'] + [0 for i in range(L-1)]
    coeff = Mlist[0]                          # coefficient of monomial
    exp = Mlist[var_position]                 # exponent of variable
    DMlist = tl.copylist(Mlist)               # list for derivative
    der_coeff = str(exp) + '('+ coeff +')'    # coeff of derivative
    DMlist[0] = ar.main(der_coeff)[0]
    DMlist[var_position] -= 1                 # reduce exponent by 1
    return DMlist

def is_constant(Mlist):    # returns True if M = [c,0,0,...]
    if Mlist[0] == '0':
        return True
    for i in range(1,len(Mlist)):
        if Mlist[i] != '0':
            return False
    return True
```

---------------------------- Sample Run ----------------------------

```
Input:
Mlist = ['7.5i',3,2,1]                        # 7.5ix^3y^2z
var_position = 1                              # x position in Mlist
print(der_mono(Mlist,var_position))
var_position = 3                              # z position in list
print(der_mono(Mlist,var_position))
```

```
Output:
['(45/2)i', 2, 2, 1]                          # (45/2)ix^2y^2z
['(15/2)i', 3, 2, 0]                          # (15/2)ix^3y^2
```

Differentiation of Polynomials

The function der_pol takes a polynomial and a variable and returns the polynomial's derivative with respect to that variable. It does so by applying der_mono to each of the monomial lists comprising that of the polynomial and then assembles the monomial lists into a polynomial list. The function then calls list2polynomial to construct the polynomial.

```
def der_pol(P,var):
    varlist = tl.get_vars(P)[0]               # the variables in P
    if var not in P: return '0'
    var_position = varlist.index(var) + 1
    Plist = main(P)[1][0]         # list for P; discard denominator 1
    DPlist = []                              # list for derivative
    for Mlist in Plist:            # run through polynomial's monos
```

8.14 Application: Partial Differentiation of Rational Functions

```
            DMlist = der_mono(Mlist,var_position) # mono derivative list
            DPlist = [DMlist] + DPlist              # attach mono list
    derivative = list2polynomial(DPlist)
    return derivative
```

---------------------------- Sample Run ----------------------------
Input:
P = 2x^3y^4z^5 + 3x^6y^7z^8
print(der_pol(P,'y'))   # differentiate with respect to y

Output:
8x^3y^3z^5+21x^6y^6z^8

Derivative of a Rational Function

The function der_quotient(num,den,var) takes a rational expression entered as numerator denominator and a variable and uses the function in the preceding section together with the quotient rule to calculate the derivative with respect to the variable var. Here's what the rule looks like using the function tl.print_fraction:

Input:
tl.print_fraction('den * der_num - num * der_den', 'den^2','')

Output:
                    den * der_num - num * der_den
                    ----
                               den^2

```
def der_quotient(R,var):
  if var not in R:
      return '0'
  if '/' not in R:
      return der_pol(R,var)       # no denominator
  num,den = R.split('/')
  der_num = der_pol(num,var)
  der_den = der_pol(den,var)
  der_quo_num = '('+ den +')('+ der_num +')-' + \
                '(('+ num +')('+ der_den + '))'
  der_quo_num = main(der_quo_num)[0]                  # clean up
  der_quo_den = '(' + den + ')^2'
  der_quo_den = main(der_quo_den)[0]
  der_quo =   '(' + der_quo_num + ')/(' + der_quo_den + ')'
  der_quo = main(der_quo)[0]                          # clean up
  return der_quo
```
---------------------------- Sample Run ----------------------------
Input:

```
R ='(7xy+3)/(5xy + 11z)'
print('partial with respect to x',der_quotient(R,'x'))
print('partial with respect to y',der_quotient(R,'y'))
print('partial with respect to z',der_quotient(R,'z'))
```

```
Output:
partial with respect to x: (77yz-15y)/(25x^2y^2+110xyz+121z^2)
partial with respect to y: (77xz-15x)/(25x^2y^2+110xyz+121z^2)
partial with respect to z: (-77xy-33)/(25x^2y^2+110xyz+121z^2)
```

Higher Order Derivatives

The function `partial_deriv(num,den,wrt,substitutions)` finds partial derivatives of arbitrary orders of a rational function of several variables. It takes a numerator and denominator, a string `wrt` of variables to differentiate with respect to, and an optional list of numerical substitutions for the variables, and returns the higher order derivative in symbolic form and a numerical value for the derivative from the substitutions.

```
def partial_derivative(R, wrt, substitutions):
    wrt = list(wrt)
    D = R
    for var in wrt:                    # get successive derivatives
        D = der_quotient(D,var)
    value = ''
    if substitutions != '':
        value = evaluate(D,substitutions,'')
```
--------------------------- Sample Run ---------------------------
```
Input:
R = '(2xy+3z)/(xy-1)'
wrt = 'yx'
subs = 'x = 1,y = 2'
D,val = partial_derivative(R, wrt, subs)
print('derivative wrt '+ wrt +':\n', D)
print('value at ' + subs + ':', val)
```

```
Output:
derivative wrt yx:
(3x^2y^2z+2x^2y^2-3z-2)/(x^4y^4-4x^3y^3+6x^2y^2-4xy+1)
value at x = 1,y = 2: 9z+6
```

8.14 Application: Partial Differentiation of Rational Functions

Tangent Plane

The function `partial_derivative` may be used to find the equation

$$z = f(a,b) + f_x(a,b)(x-a) + f_y(a,b)(y-b)$$

of the plane tangent to a surface $z = f(x,y)$ at a point $(a, b, f(a,b))$ The function `tangent_plane(num,den,a,b)` returns this equation for a rational function num/den.

```
def tangent_plane(R,a,b):
    a = str(a); b = str(b)
    subs = 'x =' + str(a) + ',y =' + str(b)
    c = evaluate(R,subs,'')
    pdx = partial_derivative(R, 'x',subs)[1]    # value at (a,b)
    pdy = partial_derivative(R, 'y',subs)[1]
    pdx = tl.add_parens(pdx)
    pdy = tl.add_parens(pdy)
    plane = c + '+' + pdx + '(x-' + a + ')+' + pdy + '(y-'+ b +')'
    plane = tl.fix_signs(plane)
    return 'z = '+ plane
```

---------------------------- Sample Run ----------------------------
```
Input:
R = '(2x^2y+3)/(3xy^3-1)'
print(tangent_plane(R,1,2))
```

```
Output:
z = 7/23+(16/529)(x-1)+(-206/529)(y-2)
```

Taylor Series in Two Variables

The Taylor series $T(x,y)$ of a function $f(x,y)$ of two variables about a point (a,b) is defined as

$$T(x,y) = \sum_{j,k} \frac{1}{j!k!} \frac{\partial^{j+k} f}{\partial^j x \, \partial^k y}(a,b)(x-a)^j (y-b)^k.$$

where the sum is taken over all pairs of nonnegative integers j, k. It is the analog of the one dimensional Taylor series developed in Sect. 6.10. To simplify notation set

$$c_{j,k} = \frac{\partial^{j+k} f}{\partial^j x \, \partial^k y}(a,b), \quad s = (x-a), \text{ and } t = (y-b),$$

where, by convention, $c_{0,0} = f(a,b)$. The first few terms of the Taylor series may then be written

$$c_{0,0} +$$
$$c_{1,0} s + c_{0,1} t +$$
$$\tfrac{1}{2!}(c_{2,0} s^2 + 2c_{1,1} st + c_{0,2} t^2) +$$
$$\tfrac{1}{3!}(c_{3,0} s^3 + 3c_{2,1} s^2 t + 3c_{1,2} st^2 + c_{0,3} t^3) +$$
$$\tfrac{1}{4!}(c_{4,0} s^4 + 4c_{3,1} s^3 t + 6c_{2,2} s^2 t^2 + 4c_{1,3} st^3 + c_{0,4} t^4)$$

Here we have separated out the zero order term, the first order terms, the second order terms, the third order terms and the fourth order terms. We denote these by S_0, S_1, S_2, S_3 and S_4, respectively. The general nth order term may be written as

$$S_n = \frac{1}{n!}\left[\binom{n}{0} c_{n,0} s^n + \binom{n}{1} c_{n,1} s^{n-1} t + \binom{n}{2} c_{n,2} s^{n-2} t^2 + \cdots \right.$$
$$\left. + \binom{n}{n-1} c_{1,n-1} st^{n-1} + \binom{n}{n} c_{0,n} t^n \right]$$
$$= \sum_{k=0}^{n} \frac{1}{(n-k)!k!} c_{n-k,k} s^{n-k} t^k$$

The first version shows the similarity of the expansion to that of the binomial theorem. The second version absorbs the $1/n!$ into the terms and is the version we use in the Python implementation below. We shall call S_n a *fixed order term*. We define the *nth order Taylor polynomial* as

$$T_n = S_0 + S_1 + \cdots + S_n.$$

Note that the degree of T_n is no more than n.

The above analysis shows that the Taylor series T is the limit of the sequence $\{T_n\}$ of Taylor polynomials. The crucial point here is that $f = T$ for suitable f, including rational functions. Thus T_n may be used as an approximation to f and its integrals, although the details are nontrivial. Here we simply construct a function that calculates the nth order Taylor polynomial of a rational function. The following function that calculates the sum S_n for arbitrary n.

```
def fixed_order_term(R,a,b,n):
    S = ''                                  # initialize fixed order term
    for k in range(n+1):                    # 0 <= k <= n
        j = n-k
        coeff = term_coeff(R,a,b,j,k)       # calculates (1/j!k!)c_j,k
        if coeff == '0' or coeff == '':
            continue
        if coeff == '1' and n != 0:
            coeff = ''                      # don't attach trivial coefficient
        if coeff == '-1':
            coeff = '-'                     # ditto
        exp_s = str(j)
        exp_t = str(k)
```

8.14 Application: Partial Differentiation of Rational Functions

```
            coeff = tl.add_parens(coeff)
            term  = coeff + 's^'+ exp_s + 't^'+ exp_t
            S = S + ' + ' + term                        # attach term
        if S == '':  # happens when partial are zero (maybe for large n)
            return ''
        # format S
        S = S.replace('s^0','')
        S = S.replace('t^0','')
        if exp_s == '1': S = S.replace('s^1','s') # skip cases s^10,...
        if exp_t == '1': S = S.replace('t^1','t')
        S = S.replace('s','(' + main('x-('+ a +')')[0] + ')')
        S = S.replace('t','(' + main('y-('+ b +')')[0] + ')')
        S = S.replace('(x-0)','x')
        S = S.replace('(y-0)','y')
        S= tl.fix_signs(S)
        return S

    def term_coeff(R,a,b,j,k):
        wrt = j*'x'+ k*'y'
        pval = partial_derivative(R,wrt,'x ='+ a + ',y = '+ b)[1]
        f = str(ma.factorial(j)*ma.factorial(k))
        coeff = ar.main('(1/('+ f +'))*(' + pval + ')')[0]
        return coeff
```

Here is the function that puts it all together. The sample run expands a 10th degree monomial about the point (.5, .3). We take $N = 10$ to get the complete (finite) Taylor series. Any $N > 10$ gives the same result. The function also checks the series by running it through main.

```
    def taylor_series(P,a,b,N):
        TS = '' # list of fixed order terms
        TS = fixed_order_term(R,a,b,0)
        for n in range(1,N+1):
            fot = fixed_order_term(R,a,b,n)
            if fot == '': continue
            TS = TS + '+' + fot
        TS = tl.fix_signs(TS)
        return TS
    ------------------------- Sample Run ------------------------
    Input:
    TS = taylor_series('x^4y^6',.5,.3,10)
    print(TS,'\n')
    # check:
    print(main(TS)[0])
    -------------------------------------------------------------
    Output:
    (729/16000000)+(729/2000000)(x+(-1/2))+(729/800000)(y+(-3/10))
```

```
+(2187/2000000)(x+(-1/2))^2+(729/100000)(x+(-1/2))(y+(-3/10))
+(243/32000)(y+(-3/10))^2
+(729/500000)(x+(-1/2))^3+(2187/100000)(x+(-1/2))^2(y+(-3/10))
+(243/4000)(x+(-1/2))(y+(-3/10))^2
+(27/800)(y+(-3/10))^3+(729/1000000)(x+(-1/2))^4
+(729/25000)(x+(-1/2))^3(y+(-3/10))
+(729/4000)(x+(-1/2))^2(y+(-3/10))^2+(27/100)(x+(-1/2))(y+(-3/10))^3
+(27/320)(y+(-3/10))^4+(729/50000)(x+(-1/2))^4(y+(-3/10))
+(243/1000)(x+(-1/2))^3(y+(-3/10))^2+(81/100)(x+(-1/2))^2(y+(-3/10))^3
+(27/40)(x+(-1/2))(y+(-3/10))^4+(9/80)(y+(-3/10))^5
+(243/2000)(x+(-1/2))^4(y+(-3/10))^2+(27/25)(x+(-1/2))^3(y+(-3/10))^3
+(81/40)(x+(-1/2))^2(y+(-3/10))^4+(9/10)(x+(-1/2))(y+(-3/10))^5
+(1/16)(y+(-3/10))^6+(27/50)(x+(-1/2))^4(y+(-3/10))^3
+(27/10)(x+(-1/2))^3(y+(-3/10))^4+(27/10)(x+(-1/2))^2(y+(-3/10))^5
+(1/2)(x+(-1/2))(y+(-3/10))^6+(27/20)(x+(-1/2))^4(y+(-3/10))^4
+(18/5)(x+(-1/2))^3(y+(-3/10))^5+(3/2)(x+(-1/2))^2(y+(-3/10))^6
+(9/5)(x+(-1/2))^4(y+(-3/10))^5+2(x+(-1/2))^3(y+(-3/10))^6
+(x+(-1/2))^4(y+(-3/10))^6

x^4y^6                                                          # check ok
-------------------------------------------------------------------
```

Linear Equations 9

In this chapter we construct the module `LinSolve.py`, which provides the complete solution of a system of linear equations with Gaussian rational coefficients. The module converts a system of equations, entered as a comma-separated string, into an augmented matrix, feeds this to a function that finds the row echelon form of the coefficient matrix, and then converts the reduced echelon form into the solution of the system. The variables of the system may be letters with or without subscripts. As usual, the letter i is reserved for complex numbers and is not allowed to be a variable name. The terms in an equation may appear in any order; their order in the augmented matrix and final solution is determined by the ordering in the list `variables`. All matrices have Gaussian rational entries. The module is headed by

```
------------------------- LinSolve.py -------------------------
import Arithmetic as ar
import Tools as tl
import MultiAlg as mu
global prod                              # for future use
global switches                          # for future use
global ops
global display
---------------------------------------------------------------
```

9.1 Matrices

An $m \times n$ *dimensional* (complex) *matrix* is a rectangular array of mn complex numbers denoted variously by

$$A = \begin{bmatrix} a_{ij} \end{bmatrix}_{m \times n} = \begin{bmatrix} a_{11} & a_{12} & \cdots & a_{1j} & \cdots & a_{1n} \\ a_{21} & a_{22} & \cdots & a_{2j} & \cdots & a_{2n} \\ \vdots & \vdots & \ddots & \vdots & \ddots & \vdots \\ a_{i1} & a_{i2} & \cdots & a_{ij} & \cdots & a_{in} \\ \vdots & \vdots & \ddots & \vdots & \ddots & \vdots \\ a_{m1} & a_{m2} & \cdots & a_{mj} & \cdots & a_{mn} \end{bmatrix}. \qquad (9.1)$$

The number a_{ij} in row i and column j is called the (i, j) *entry* of A. For example, the $(2, 3)$ entry of the 2×4 matrix $\begin{bmatrix} 1 & 2 & 3 & 4 \\ 5 & 6 & 7 & 8 \end{bmatrix}$ is the number 7. An $n \times n$ matrix is said to have *size n*. Two matrices A and B are said to be *equal* if they have the same dimensions and equal entries: $a_{ij} = b_{ij}$ for all indices i and j.

The notation in (9.1) is standard: indices start at 1. However, in the programs that follow, matrices are stored as double lists and hence have indices that start at zero. One must therefore be careful in writing programs that involve matrices, keeping in mind the dual notation. For example, for the case $m = 3$ and $n = 2$ the above matrix is stored as a list with entries

```
A = [ [A[0][0], A[0][1]], [A[1][0], A[1][1]], [A[2][0], A[2][1]] ].
```

Here, the list [A[0][0],A[0][1]] is the first row, [A[1][0],A[1][1]] the second row, and [A[2][0],A[2][1]] the third row. The connection with (9.1) is given by the conversion formula,

$$a_{ij} = A[i-1][j-1]|, \text{ where } i, j \geq 1.$$

All matrices in the text have entries that are Gaussian rational numbers or arithmetic expressions of these, all written as strings. The module Arithmetic handles these easily, producing exact fractional values. One can use ar.decimal_approx to produce decimal approximations to any desired degree of accuracy.

9.2 Systems of Linear Equations

An $m \times n$ *system of linear equations* is a collection of m equations in n unknowns of the form

$$\begin{aligned} a_{11}x_1 + a_{12}x_2 + \cdots + a_{1n}x_n &= b_1 \\ a_{21}x_1 + a_{22}x_2 + \cdots + a_{2n}x_n &= b_2 \\ &\vdots \\ a_{m1}x_1 + a_{m2}x_2 + \cdots + a_{mn}x_n &= b_m. \end{aligned} \quad (9.2)$$

The symbols x_j are variables and the symbols a_{ij}, b_i represent constants. In general these are complex numbers but for our purposes are restricted to be Gaussian rationals. A *solution* of the system (9.2) is a set of values of the variables $x_1, x_2, \ldots, x_n$ that satisfies the equations. As we shall see, a system may have no solutions, a unique solution, or infinitely many solutions.

All of the essential information in the above system may be represented by its *augmented matrix*[1]:

$$\begin{bmatrix} a_{11} & a_{12} & \cdots & a_{1n} & b_1 \\ a_{21} & a_{22} & \cdots & a_{2n} & b_2 \\ \vdots & \vdots & \ddots & \vdots & \vdots \\ a_{m1} & a_{m2} & \cdots & a_{mn} & b_n \end{bmatrix}$$

Thus to manipulate the equations of a system one may use the augmented matrix and manipulate the rows instead.

9.3 The Gauss-Jordan Method

The Gauss-Jordan algorithm provides a way to solve a system of linear equations, or else determines that the system has no solution. The idea is to transform the system into one that has precisely the same solutions as the original system but is trivial to solve. The transformation to a simpler system is accomplished by a sequence of *equation operations* of the following form:

(1) Interchange two equations.
(2) Multiply an equation by a nonzero number.
(3) Add a multiple of one equation to another.

---

[1] The only purpose of the vertical bar in the matrix is to emphasize that the matrix arose from a system of equations. The bar is frequently omitted if no confusion can arise.

The important thing here is that these operations do not change solutions: any solution of the original system is also a solution of a system obtained by performing these operations and vice versa.

Equation operations may be expressed as *row operations* on the augmented matrix of the system. The above operations then take the following form:

(1) Interchange two rows.
(2) Multiply a row by a nonzero number.
(3) Add a multiple of one row to another row.

The following shorthand notation will be used in the examples to indicate these operations:

Type 1   ra <-> rb   interchange rows a and b.
Type 2   (t)ra   multiply row a by a nonzero number t.
Type 3   (t)ra+rb   add t times row a to another row b.

In the last operation it is row b that is modified; row a is unchanged.

Here's an example that shows how these operations are typically used to obtain the solutions of a system:

Example 9.1

$$\begin{aligned} 4x_1 + 5x_2 + 6x_3 &= 12, \\ x_1 + 2x_2 + 3x_3 &= 9, \\ 7x_1 + 8x_2 + 9x_3 &= 15 \end{aligned} \tag{9.3}$$

We apply the following operations to the augmented matrix the system:

$$\begin{bmatrix} 4 & 5 & 6 & 12 \\ 1 & 2 & 3 & 9 \\ 7 & 8 & 9 & 15 \end{bmatrix} \xrightarrow{r1<->r2} \begin{bmatrix} 1 & 2 & 3 & 9 \\ 4 & 5 & 6 & 12 \\ 7 & 8 & 9 & 15 \end{bmatrix}$$

$$\xrightarrow[(-7)r1+r3]{(-4)r1+r2} \begin{bmatrix} 1 & 2 & 3 & 9 \\ 0 & -3 & -6 & -24 \\ 0 & -6 & -12 & -48 \end{bmatrix} \xrightarrow{(-1/3)r2}$$

$$\begin{bmatrix} 1 & 2 & 3 & 9 \\ 0 & 1 & 2 & 8 \\ 0 & -6 & -12 & -48 \end{bmatrix} \xrightarrow[(6)r2+r3]{(-2)r2+r1} \begin{bmatrix} 1 & 0 & -1 & -7 \\ 0 & 1 & 2 & 8 \\ 0 & 0 & 0 & 0 \end{bmatrix}.$$

The last matrix is the augmented matrix of the system

$$\begin{aligned} x_1 \quad - x_3 &= -7, \\ x_2 + 2x_3 &= 8 \end{aligned} \tag{9.4}$$

which has solutions

$$x_1 = -7 + x_3, \quad x_2 = 8 - 2x_3, \quad x_3 \text{ arbitrary.} \tag{9.5}$$

9.3 The Gauss-Jordan Method

The system (9.4) therefore has infinitely many solutions, one for each value of the parameter x_3. Moreover, as observed earlier, row operations don't modify solutions, so every solution of the original system is of the form (9.5). Conversely, every solution of this form is a solution of the original system. This is so because we can reverse the operations that led to (9.4):

$$\begin{bmatrix} 1 & 0 & -1 & -7 \\ 0 & 1 & 2 & 8 \\ 0 & 0 & 0 & 0 \end{bmatrix} \xrightarrow[(-6)r2+r3]{(2)r2+r1} \begin{bmatrix} 1 & 2 & 3 & 9 \\ 0 & 1 & 2 & 8 \\ 0 & -6 & -12 & -48 \end{bmatrix}$$

$$\xrightarrow{(-3)r2} \begin{bmatrix} 1 & 2 & 3 & 9 \\ 0 & -3 & -6 & -24 \\ 0 & -6 & -12 & -48 \end{bmatrix} \xrightarrow[(7)r1+r3]{(4)r1+r2}$$

$$\begin{bmatrix} 1 & 2 & 3 & 9 \\ 4 & 5 & 6 & 12 \\ 7 & 8 & 9 & 15 \end{bmatrix} \xrightarrow{r1<->r2} \begin{bmatrix} 4 & 5 & 6 & 12 \\ 1 & 2 & 3 & 9 \\ 7 & 8 & 9 & 15 \end{bmatrix}$$

It now follows that the original linear system has precisely the solutions given in (9.5). ◊

The above example illustrates the *Gauss-Jordan method*: a sequence of row operations that transform a matrix into one with the following properties:

- All zero rows (rows with only zeros) are below all nonzero rows (rows with at least one non-zero).
- The first nonzero entry (called the *leading entry*) in a nonzero row is 1.
- The leading entry in one row is to the left of all leading entries below it.
- All entries above and below a leading entry are zero.

A matrix with the above properties is said to be in *reduced row echelon form*. For example, the first matrix below is in reduced row echelon form, the second is not. For emphasis we have enclosed the leading entries in the first matrix in rectangles.

$$\begin{bmatrix} \boxed{1} & 3 & 0 & 0 & 0 \\ 0 & 0 & \boxed{1} & 0 & 0 \\ 0 & 0 & 0 & \boxed{1} & 2 \\ 0 & 0 & 0 & 0 & 0 \end{bmatrix} \quad \begin{bmatrix} 1 & 3 & 5 & 1 \\ 0 & 0 & 7 & 2 \\ 0 & 0 & 0 & 0 \\ 0 & 0 & 0 & 3 \end{bmatrix}.$$

Any matrix may be transformed into reduced row echelon form by a sequence of row operations. In particular, this can be done for the augmented matrix of a system of linear equations. Since the system of linear equations corresponding to a matrix in reduced row echelon form is essentially trivial, we now have a way of finding the solutions of any system (or determining that the system has no solutions).

The procedure used to transform a matrix A into row reduced echelon form is described in the following steps. (We assume that not every entry of A is zero.)

(a) Find the leftmost column that has at least one nonzero entry. This is called a *pivot column*. The *pivot position* is at the top of the column in what is called the *pivot row*.
(b) Choose a nonzero entry in the pivot column.
(c) Use a type 1 row operation on the matrix to move the entry to the pivot position.
(d) Use a type 2 row operation to put a 1 in the pivot position.
(e) Use type 3 row operations to put zeros in all but the pivot position of the pivot column.
(f) Find the leftmost non-zero column in the matrix consisting of the rows below the last pivot row and apply steps (b)–(e). Continue the process until there are no more rows left to modify.

We shall call the process of placing a one in the pivot position of a column and zeros elsewhere *clearing the column*.

Here are some additional examples of the Gauss-Jordan method of solving systems of linear equations.

Example 9.2 Consider the following system obtained from Example 9.1 by changing the coefficient of x_1 in the second equation from 1 to 2:

$$4x_1 + 5x_2 + 6x_3 = 12,$$
$$2x_1 + 2x_2 + 3x_3 = 9,$$
$$7x_1 + 8x_2 + 9x_3 = 15$$

Here are row operations that reduce the augmented matrix to row echelon form. Again, for emphasis we have enclosed in a rectangle the entry used to clear a column.

$$\begin{bmatrix} 4 & 5 & 6 & 12 \\ 2 & 2 & 3 & 9 \\ 7 & 8 & 9 & 15 \end{bmatrix} \xrightarrow{r1<->r2} \begin{bmatrix} 2 & 2 & 3 & 9 \\ 4 & 5 & 6 & 12 \\ 7 & 8 & 9 & 15 \end{bmatrix} \xrightarrow{(1/2)r1} \begin{bmatrix} \boxed{1} & 1 & \frac{3}{2} & \frac{9}{2} \\ 4 & 5 & 6 & 12 \\ 7 & 8 & 9 & 15 \end{bmatrix}$$

$$\xrightarrow[(-7)r1+r3]{(-4)r1+r2} \begin{bmatrix} 1 & 1 & \frac{3}{2} & \frac{9}{2} \\ 0 & \boxed{1} & 0 & -6 \\ 0 & 1 & -\frac{3}{2} & -\frac{33}{2} \end{bmatrix} \xrightarrow[(-1)r2+r3]{(-1)r2+r1} \begin{bmatrix} 1 & 0 & \frac{3}{2} & \frac{21}{2} \\ 0 & 1 & 0 & -6 \\ 0 & 0 & -\frac{3}{2} & -\frac{21}{2} \end{bmatrix}$$

$$\xrightarrow{(-2/3)r3} \begin{bmatrix} 1 & 0 & \frac{3}{2} & \frac{21}{2} \\ 0 & 1 & 0 & -6 \\ 0 & 0 & \boxed{1} & 7 \end{bmatrix} \xrightarrow{(-3/2)r3+r1} \begin{bmatrix} 1 & 0 & 0 & 0 \\ 0 & 1 & 0 & -6 \\ 0 & 0 & 1 & 7 \end{bmatrix}$$

The last matrix is the augmented matrix of the system

$$x_1 \quad\quad\quad = 0,$$
$$x_2 \quad\quad = -6,$$
$$x_3 = 7$$

which is therefore the solution of the original system. ◊

9.4 Row Operations in List Form

Example 9.3 Consider the system

$$x_1 + 2x_2 + 3x_3 + 4x_4 = 5,$$
$$6x_1 + 7x_2 + 8x_3 + 9x_4 = 10,$$
$$11x_1 + 12x_2 + 13x_3 + 14x_4 = 15,$$
$$16x_1 + 17x_2 + 18x_3 + 19x_4 = 20$$

Clearing column 1 and then column 2 in the augmented matrix we have

$$\begin{bmatrix} 1 & 2 & 3 & 4 & 5 \\ 6 & 7 & 8 & 9 & 10 \\ 11 & 12 & 13 & 14 & 15 \\ 16 & 17 & 18 & 19 & 20 \end{bmatrix} \longrightarrow \begin{bmatrix} 1 & 2 & 3 & 4 & 5 \\ 0 & -5 & -10 & -15 & -20 \\ 0 & -10 & -20 & -30 & -40 \\ 0 & -15 & -30 & -45 & -60 \end{bmatrix}$$

$$\longrightarrow \begin{bmatrix} 1 & 0 & -1 & -2 & -3 \\ 0 & 1 & 2 & 3 & 4 \\ 0 & 0 & 0 & 0 & 0 \\ 0 & 0 & 0 & 0 & 0 \end{bmatrix}.$$

The last matrix corresponds to the equivalent system

$$x_1 \quad - x_3 - 2x_4 = -3,$$
$$x_2 + 2x_3 + 3x_4 = 4$$

which has solutions

$$x_1 = x_3 + 2x_4 - 3, \quad x_2 = -2x_3 - 3x_4 + 4, \quad x_3, x_4 \text{ arbitrary}.$$

Thus the set of solutions is described by two parameters, x_3 and x_4. ◇

It is easy to construct systems that have no solutions, for example,

$$x_1 + x_2 = 1,$$
$$x_1 + x_2 = 2$$

Such systems are said to be *inconsistent*. Note that the last row of the reduced augmented matrix for the system is [0,0,1], which is characteristic of inconsistent systems.

9.4 Row Operations in List Form

The symbolic strings used in the above examples that denote the row operations are used in the module for concise, clear display. For internal use it is convenient to have a description of the row operation in terms of a list. Such a list has the following entries: The first is the operation type number; the remaining entries give the specifics in terms of rows and scalars. The following examples illustrate the idea.

(1) [1,'', 3, 4]

A type 1 operation that switches rows 3 and 4 of a matrix. The symbolic version is 'r3 <-> r4'.

(2) [2,(1.2 + 3.4i),5,'']

A type 2 operation that multiplies row 5 by the scalar (1.2 + 3.4i). The symbolic version is '(1.2 + 3.4i)r5'.

(3) [3,(12.3/45.6),1,3]

A type 3 operation that adds to row 3 the result of multiplying row 1 by (12.3/45.6). The symbolic version is '(12.3/45.6)r1+r3'.

In (1) no scalar in used, so a null string is put in that position. Similarly, in (2) only one row is needed, so a null string is placed in the last position.

For the upcoming functions we shall refer to the list version of an operation as simply an op. To distinguish between the two ways of referring to an operation, we refer the string version as an opsymbol.

The following functions convert from one representation of a row operation to the other, that is, from opsymbol to op and vice-versa.

```
def sym2op(opsym):                          # expand string opsym into list op
    opsym = opsym.replace(' ','')
    if '<' in opsym:                                         # a<->b
        a,b = opsym.split('<->')                             # no scalar
        a = a.replace('r','')                     # remove the 'r' prefix
        b = b.replace('r','')
        op = [1,'',int(a),int(b)]                 # type 1: switch the rows
        return op
    # extract scalar t from (t)a or (t)a+b:
    right_most_paren = opsym.rfind(')')                      # index of ')'
    s = opsym[:right_most_paren+1]            # extract scalar and parens
    the_rest = opsym[right_most_paren+1:]     # stuff after right paren
    the_rest = the_rest.replace('r','')       # remove the 'r' prefix
    s = '(' + ar.main(s)[0] + ')'                            # simplify
    if '+' in the_rest:                                      # '(scalar)a+b'
        a, b = the_rest.split('+')                           # extract the rows
        op = [3,s,int(a),int(b)]                             # type 3
    else:                                     # remaining case '(scalar)a'
        op = [2,s,int(the_rest), '']                         # type 2
    return op

------------------------------ Sample Run ------------------------------
Input:
print('symbolic form:       ','list form:')
print('r2<->r3              ',sym2op('r2<->r3'))
print('(4.2-7/2i + 3.2i)r5  ',sym2op('(4.2-7/2i + 3.2i)r5'))
print('(4.6/5.78)r7 + r11   ',sym2op('(4.6/5.78)r7 + r11'),'\n')
------------------------------------------------------------------------
Output:
```

```
symbolic form:      list form:
r2<->r3             [1, '', 2, 3]
(4.2-7/2i + 3.2i)r5 [2, '(21/5+(67/10)i)', 5, '']
(4.6/5.78)r7 + r11  [3, '(230/289)', 7, 11]
```

```
def op2sym(op):
    if op[0] == 1:                                      # [1,'',a,b]
        a = op[2]; b = op[3]                            # adjust row numbers
        op = 'r' + str(a) + '<->' + 'r' + str(b)        # ra<->rb
    elif op[0] == 2:                                    # [2,t,a,'']
        scalar = op[1]; a = op[2]
        op = scalar + 'r' + str(a)                            (t)ra
    else:                                               # [3,t,a,b]
        scalar = op[1]
        a = op[2]
        b = op[3]
        op = scalar +'r'+ str(a) + '+' +'r'+ str(b)     # (t)ra + rb
    return op
```

9.5 Implementing Row Operations

The function row_op_calc(op,A) takes a matrix A and an operation op and applies the operation to the matrix.

```
def row_op_calc(op, A):
    nrows, ncols = len(A), len(A[0])
    optype = op[0]; scalar = op[1]
    B = tl.copylist(A)
    if optype == 1:                             # switch rowa and rowb
        for j in range(ncols):
            rowa = op[2]-1; rowb = op[3]-1      # list index convention
            temp = B[rowa][j]                           # store rowa entry
            B[rowa][j] = B[rowb][j]             # copy rowb entry into rowa
            B[rowb][j] = temp                   # copy old rowa entry into rowb

    elif optype == 2:                                   # scalar mult
        rowa = op[2]-1
        for j in range(ncols):                          # scalar times row
            a = '(' + scalar + ')(' + B[rowa][j] + ')'
            B[rowa][j] = ar.main(a)[0]

    elif optype == 3:                           # add scalar*rowa to rowb
        rowa = op[2]-1; rowb = op[3]-1
        for j in range(ncols):
```

```
                a = '(' + scalar + ')(' + B[rowa][j] + ')+' + B[rowb][j]
                a = tl.fix_signs(a)
                B[rowb][j] = ar.main(a)[0]
    return B
```

The only purpose of following function is to illustrate the steps in the evolution of a matrix to its row reduced form. The sample run is Example 9.2. (Output is written horizontally to save space.)

```
def run_ops(opsyms,A):                  # apply a list of opsyms to A
    print('        A')                                       # label
    tl.format_print(A,2,'right'); print('\n')
    B = tl.copylist(A)
    for opsym in opsyms:
        op = sym2op(opsym)              # convert operation to list form
        B = row_op_calc(op,B)
        print('   ',opsym)                                   # label
        tl.format_print(B,2,'right')                  # check progress
        print('\n')
    return B

--------------------------- Sample Run ---------------------------
Input:
opsyms = 'r1<->r2,(1/2)r1,(-7)r1+r3,(-4)r1+r2,\
          (-1)r2+r3,(-1)r2+r1,(-2/3)r3,(-3/2)r3+r1'
opsyms_list = opsyms.split(',')
A ='4,5,6,12; 2,2,3,9; 7,8,9,15'
A = tl.string2table(A)
print(run_ops(opsyms_list,A))
------------------------------------------------------------------
Output:
       A                r1<->r2             (1/2)r1               (-7)r1+r3
  4  5  6  12        2  2  3   9        1  1  3/2  9/2        1  1   3/2    9/2
  2  2  3   9        4  5  6  12        4  5   6    12        4  5    6     12
  7  8  9  15        7  8  9  15        7  8   9    15        0  1  -3/2  -33/2

    (-4)r1+r2              (-1)r2+r3              (-1)r2+r1
  1  1   3/2    9/2     1  1   3/2    9/2      1  0   3/2   21/2
  0  1    0     -6      0  1    0     -6       0  1    0     -6
  0  1  -3/2  -33/2     0  0  -3/2  -21/2      0  0  -3/2  -21/2

    (-2/3)r3              (-3/2)r3+r1
  1  0  3/2  21/2      1  0  0   0
  0  1   0    -6       0  1  0  -6
  0  0   1     7       0  0  1   7
------------------------------------------------------------------
```

9.6 Row Echelon Form in Python

The function `row_echelon(A)` below follows the row-echelon algorithm by finding at each step a pivot column and a non-zero entry in the column (pivot entry), moving the row containing this entry up to become a lead row, and then clearing the pivot column using the pivot entry. The main work is carried out by the function `row_op_calc`. The job of `row_echelon` is to assign the appropriate row operation.

The function `row_echelon` also keeps track of the number of row switches in the variable `switches` and calculates the product `prod` of the matrix entries whose reciprocals are the scalar multipliers in type 2 operations during the pivoting process. These variables will be used in the section on determinants in Chap. 12 and may be ignored for now.

```
def row_echelon(A):
    global prod, switches              # for future use
    global ops
    prod = '1'; switches = 0           # initialize (later chapter)
    ops = []
    nrows, ncols = len(A), len(A[0])
    toprow = 0; row = 0; col = 0       # begin here
    B = tl.copylist(A)                 # don't change A
    while toprow < nrows and col < ncols:   # find pivot columns
        row = toprow                   # potential leading entry row
        # search in col and below toprow for row with entry != 0:
        while row < nrows and B[row][col] == '0':
            row += 1       # keep going until found entry!=0 in col
        if row < nrows:
            if B[row][col] != '0':     # if found entry != 0
                # update prod and move row to first
                prod = '('+ prod +')('+ B[row][col] +')'
                prod = ar.main(prod)[0]
                if row != toprow:
                    switches += 1                  # update
                    # switch row and toprow
                    op = [1,'',row+1,toprow+1]  # op row convention
                    ops.append(op)                 # keep a record
                    B = row_op_calc(op, B)            # switched
                B = clear_col(B, toprow, col)   # toprow lead row
                toprow = toprow + 1                # next toprow
        col = col + 1      # next col to search for next pivot entry
    return B, ops # return echelon form and operations
```

--------------------------- Sample Run ---------------------------
```
Input:
A = tl.string2table('1/3,-11,7-4i,-10;-i,-2,7.8+3.8i,-15;1,i,1,-20')
print('             A')
tl.format_print(A,2,'right'); print('')
B, ops = row_echelon(A)
print('             reduced')
tl.format_print(B,2,'right')
```

```
Output:
            A
  1/3   -11        7-4i    -10
   -i    -2    7.8+3.8i    -15
    1     i           1    -20

            reduced
   1   0   0   -354753/21661+(45496/21661)i
   0   1   0   -103133/64983+(16747/21661)i
   0   0   1    -61720/21661-(33355/64983)i
```

The function `clear_col` uses the pivot entry `B[toprow][col]` to clear the column

```
def clear_col(B, toprow, col):
    global ops
    nrows, ncols = len(B), len(B[0])
    pivot_entry = B[toprow][col]          # use entry to clear col
    scalar = '1/(' + pivot_entry + ')'
    op = [2, scalar, toprow+1, '']        # divide toprow by pivot_entry
    ops.append(op)                        # save for op record
    B = row_op_calc(op, B)                # make the division
    # replace each row by 'row -B[row, col]*toprow:
    for row in range(nrows):
        if row == toprow or B[row][col] == '0':
            continue           # skip toprow and zero entry multiplier
        simplify = ar.main('-' + '(' + B[row][col] + ')')[0]
        op = [3, '(' + simplify + ')', toprow+1, row+1]
        B = row_op_calc(op, B)            # do the operation on B
        ops.append(op)                    # save for op record
    return B
```

9.7 Reduced Column Echelon Form

The algorithm used to row reduce a matrix A may be naturally modified to produce the analogous column reduced form. This amounts to applying `row_echelon` to the matrix B obtained from A by converting columns of A into rows and then converting the columns of the row echelon form into rows. The matrix B is called the *transpose* of A and is discussed further in Chap. 10 in connection with matrix algebra.

Here is the function that produces the transpose. (Output in sample run is placed horizontally to save space.)

9.7 Reduced Column Echelon Form

```
def transpose(A):
    nrows, ncols = len(A),len(A[0])
    T = [['' for j in range(nrows)] for i in range(ncols)]
    for j in range(nrows):
        for i in range(ncols):    # make jth col of T the jth row of A
            T[i][j] = A[j][i]
    return T
```

------------------------- Sample Run -------------------------
```
Input:
A = tl.string2table('1,2,3,4;5,6,7,8;9,10,11,12')
T = transpose(A)
print('        A ')
tl.format_print(A,2,'right')
print('\n')
print(' A transpose ')
tl.format_print(T,2,'right')
```

Output:
```
        A                       A transpose
   1    2    3    4                1   5   9
   5    6    7    8                2   6  10
   9   10   11   12                3   7  11
                                   4   8  12
```

The function `col_echelon(A)` returns the column reduced echelon form. The sample run includes an interesting test. It may be shown that the special form of the last two matrices and their equality holds generally.

```
def col_echelon(A):
    T = transpose(A)            # rows to columns and columns to rows
    return transpose(row_echelon(T)[0])
```

------------------------- Sample Run -------------------------
```
Input:
A = tl.string2mat('1,-2,3,-4,5;-6,7,-8,9,-10;11,-12,13,-14,15')
print('A:')
tl.format_print(A,2,'right');print('\n')

RA = row_echelon(A)[0]
print('row echelon of A:')
tl.format_print(RA,2,'right');print('\n')

CA = col_echelon(A)
print('col echelon of A:')
tl.format_print(CA,2,'right');print('\n')

CRA = col_echelon(RA)
print('col echelon of row echelon A:')
```

```
        tl.format_print(CRA,2,'right');print('\n')

        RCA = row_echelon(CA)[0]
        print('row echelon of col echelon A:')
        tl.format_print(RCA,2,'right')
```

```
Output:
A:
    1   -2    3   -4    5
   -6    7   -8    9  -10
   11  -12   13  -14   15

row echelon of A:
    1    0   -1    2   -3
    0    1   -2    3   -4
    0    0    0    0    0

col echelon of A:
    1    0    0    0    0
    0    1    0    0    0
   -1   -2    0    0    0

col echelon of row echelon A:
    1    0    0    0    0
    0    1    0    0    0
    0    0    0    0    0

row echelon of col echelon A:
    1    0    0    0    0
    0    1    0    0    0
    0    0    0    0    0
```

9.8 Linsolve

The function linsolve launches the process of solving a system of linear equations. The function takes as input either a string of equations or an augmented matrix and returns the corresponding solution of the system in the form of a list. The print statements show the intermediate steps. These are activated by setting the global variable display to True. Deactivate them by setting the variable to False.

```
    def linsolve(equations,augmat,letter,display):
        global varlist
        if equations != []:         # if string of equations were entered
            eqnlist = equations.split(',')      # convert them to lists
            varlist = get_system_variables(eqnlist)
            augmat = get_augmat(eqnlist,varlist) # make augmented matrix
```

9.8 Linsolve

```
        else:                           # augment matrix was entered
            varlist = make_variables(len(augmat[0])-1,letter)

        if display:
            print('augmat:')
            tl.format_print(augmat,4, 'right')
            print('\n')

        reduced = row_echelon(augmat)[0]        # get row reduced form

        if display:
            print('reduced augmat:')
            tl.format_print(reduced,4, 'right')
            print('\n')
        sol_list = get_solution_list(reduced,varlist)
        return sol_list
```

Here are four sample runs with `display` set to True. In the first three runs the matrix variable `augmat` is set to the empty [], since equation lists are provided. In the last example the `equations` variable is set to the null string, since the augmented matrix is given. The first example has a unique solution.

```
Input:
equations =   '(1+i)x2 +   (2+i)x3   +   x4   + x5 = 1/2, \
               x1                  +   x3    +   x4   + x5 = 2/3, \
               x1   +   x2                   +   x4   + x5 = 3/4, \
               x1   +   x2   +   x3                  + x5 = 4/5, \
               x1   +   x2   +   x3    +   x4             = 5/6'
print(linsolve(equations,[],'',True))

Output:
augmat:
    0      (1+i)      (2+i)       1      1      1/2
    1        0          1         1      1      2/3
    1        1          0         1      1      3/4
    1        1          1         0      1      4/5
    1        1          1         1      0      5/6

reduced:
    1    0    0    0    0    309/580+(91/580)i
    0    1    0    0    0      5/29-(91/1740)i
    0    0    1    0    0    31/348-(91/1740)i
    0    0    0    1    0    17/435-(91/1740)i
    0    0    0    0    1     1/174-(91/1740)i

['x1 = 309/580+(91/580)i', 'x2 = 5/29-(91/1740)i', \
 'x3 = 31/348-(91/1740)i', 'x4 = 17/435-(91/1740)i', \
 'x5 = 1/174-(91/1740)i']
```

The second example has infinitely many solutions, one for each set of values of the parameters x3,x4,x5.

```
------------------------------------------------------------------
Input:
equations = ' x1  +   2x2  +   3x3  +  4x4  +  5x5  = 6,\
              7x1  +   8x2  +   9x3  + 10x4  + 11x5 = 12, \
             13x1  +  14x2  +  15x3  + 16x4  + 17x5 = 18'
print(linsolve(equations,[],'',True))
------------------------------------------------------------------
Output:
augmat:
    1    2    3    4    5    6
    7    8    9   10   11   12
   13   14   15   16   17   18

reduced:
    1    0   -1   -2   -3   -4
    0    1    2    3    4    5
    0    0    0    0    0    0

['x1=3x5+2x4+x3-4', 'x2=-4x5-3x4-2x3+5', 'x3', 'x4', 'x5']
------------------------------------------------------------------
```

The third example has no solutions, as evidenced by the third row of the reduced matrix.

```
------------------------------------------------------------------
Input:
equations = ' x1  +   2x2  +   3x3  +  4x4  +  5x5 = 6,  \
              7x1  +   8x2  +   9x3  + 10x4  + 11x5 = 12, \
             13x1  +  14x2  +  15x3  + 16x4  + 17x5 = 18, \
             19x1  +  20x2  +  21x3  + 22x4  + 23x5 = 23'
print(linsolve(equations,[],'',True))
------------------------------------------------------------------
Output:
augmat:
    1    2    3    4    5    6
    7    8    9   10   11   12
   13   14   15   16   17   18
   19   20   21   22   23   23

reduced:
    1    0   -1   -2   -3    0
    0    1    2    3    4    0
    0    0    0    0    0    1       # system inconsistent
    0    0    0    0    0    0

[' ']                                 # no solution
------------------------------------------------------------------
```

9.9 Setting Up the Variables

In the final example an augmented matrix and the letter 'x' are given as inputs. The program returns a unique solution.

```
------------------------- Sample Run -------------------------
Input:
a = '11,16,6;12,17,7;13,18,8;14,19,9;15,20,10'
print(linsolve([],tl.string2table(a),'x',))
--------------------------------------------------------------
Output:
augmat:
    11    16    6
    12    17    7
    13    18    8
    14    19    9
    15    20   10

reduced:
    1    0    2
    0    1   -1
    0    0    0
    0    0    0
    0    0    0

['x1=2', 'x2=-1']
--------------------------------------------------------------
```

9.9 Setting Up the Variables

As noted earlier, the function linsolve takes either a string of equations or an augmented matrix. If the former, the program calls get_system_variables, which scans the equations for the variable names. If the latter, the program calls make_variables, which generates a standard set of subscripted variables with the base letter the user's choice. Here is the code.

```
--------------------------------------------------------------
def get_system_variables(equations):     # scan eqns for variables
    variables = []
    for eqn in equations:
        eqn = eqn.replace(' ','')
        variables += tl.get_vars(eqn)[0]
    return sorted(list(set(variables)))    # remove duplicates, sort

def make_variables(dim,letter):    # letter for subscripted variables
    varlist = []
    for n in range(1,dim+1):              # print a label for letter
        varlist.append(letter+ str(n))
    return varlist
--------------------------------------------------------------
```

9.10 Creating the Augmented Matrix

The function `get_augmat` creates the augmented matrix for the system for the case that equations are the input. It does so row by row, calling `eqn2row(eqn)` for each equation in the system.

```
def get_augmat(eqnlist,varlist):
    augmat = []
    for eqn in eqnlist:
        row = eqn2row(eqn,varlist)
        augmat.append(row)
    return augmat
```

The function `eqn2row(eqn)` takes an equation and generates its row in the augmented matrix. The order of the row entries is determined by the order of the variables in `varlist` and not necessarily their order in the equation. The order of the rows is the same as the order of the equations. The preliminary step `insert_delimiters(eqn)` separates out the terms by enclosing variables with commas, allowing easy extraction variables from the expression. Print statements follow the progress. Comment these out if desired. For added flexibility we allow a constant on the left side of an entry equation, as in $x + 2y + 3z + 4 = 5$. The only restriction is that the constant be the last term on the left side.

```
def insert_delimiters(eqn):              # place commas around variables
    letters = tl.letters
    eqn = tl.attach_missing_coeff(eqn,tl.letters)  # for missing '1'
    for letter in letters:               # place comma before letter
        eqn = eqn.replace(letter,','+letter)
    i = 0
    while i < len(eqn):
        if eqn[i] not in letters:        # skip non letters
            i += 1; continue
        start = i+1                      # after variable
        end = tl.movepast(eqn,start,'1234567890')  # past subscript
        eqn = tl.insert_string(eqn,',',end)  # at position end
        i = end
    return eqn

def eqn2row(eqn,varlist):                # convert equation to augmat row
    eqn = insert_delimiters(eqn)         # enclose variables with commas
    print('delimiters  ',eqn,'\n')       # for observation
    components = eqn.split(',')          # split off vars and coeffs
    print('components:  ',components)    # for observation
    C = len(components)
    last = components[C-1]
    if last[0] != '=':                   # place constant if any on right
        last = last.split('=')
        new_right_side = last[1] + '-(' + last[0] + ')'
```

```
                new_right_side = ar.main(new_right_side)[0]
                components[C-1]= new_right_side
        V = len(varlist)
        components[C-1] = components[C-1].replace('=','')   # right side
        row = ['0' for k in range(V)]         # initialize row with zeros
        for i in range(V):            # run through variables in varlist order
            var = varlist[i]                  # var in position i in varlist
            for j in range(0,C-2,2):    # 2-step iteration;stop before '='
                c = components[j]                           # coefficient
                v = components[j+1]                         # variable
                if var == v:
                    row[i] = c    # put coeff of var in position i of row
        row.append(components[C-1])   # put last component on right side
        for i in range(len(row)):
            row[i] = tl.fix_signs(row[i])
        return row

        ------------------------- Sample Run -------------------------
        Input:
        varlist = ['x1','x2','x3','x4','u','v']    # entry order in augmat
        eqn = '5v-3x4+5x1-4x2+11x3-9u+8=-7'

        print('row:           ',eqn2row(eqn,varlist))
        -------------------------------------------------------------
        Output:
        delimiters: 5,v,-3,x4,+5,x1,-4,x2,+11,x3,-9,u,+8=-7
        components: ['5', 'v', '-3', 'x4', '+5', 'x1', '-4', 'x2', '+11',\
                    'x3', '-9', 'u', '+8=-7']
        row:            ['5', '-4', '11', '-3', '-9', '5', '-15']
        -------------------------------------------------------------
```

9.11 Generating the Solutions

The function get_solution_list takes the reduced echelon form of the augmented matrix and the variable and outputs the solutions. The function translates a row of the reduced matrix into an equation by solving for the leading variable. For example, the row ['0','1','2', '3', '4'] is translated into 'x2=4-2x3-3x4'.

```
    -------------------------------------------------------------
    def get_solution_list(reduced,varlist):
        if has_zeros_one_row(reduced):
            return []                          # no solution
        sol_list = tl.copylist(varlist)        # initialize
        L = len(reduced)                       # number of rows
        M = len(reduced[0])                    # number of columns
        for k in range(L):                     # run throught rows
            if is_zero_row(reduced[k]): break
            right_side = reduced[k][M-1]   # last entry in row
```

```
            idx = reduced[k].index('1')   # column of leading entry '1'
            v = varlist[idx]              # corresponding variable
            # subtract other variables from right side:
            for j in range(idx+1,M-1):
                right_side = right_side + \
                             '-('+reduced[k][j]+')' + varlist[j]
                right_side = tl.fix_signs(right_side)
            right_side = mu.main(right_side)[0]
            sol_list[idx] = v + '='  + right_side
    return sol_list

def is_zero_row(row):
    L = len(row)
    for entry in row:                     # zero row if L is reduced to 0
        if entry == '0': L -= 1
    return L == 0
```

The function has_zeros_one_row(R) checks if the reduced matrix R has a row of the form ['0','0',...,'0','1'], signalling an inconsistent equation.

```
def has_zeros_one_row(R):
    for row in R:
        if row[len(row)-1] == '0':
            return False
        num_zeros = 0
        for j in range(len(row)-1):
            if row[j] == '0': num_zeros += 1
        if num_zeros == len(row)-1:                          '0,0,...0,1'
            return True
    return False
```

9.12 Checking the Solution

The function check_solution(equations,solutions val) takes the original set of equations, the solutions generated by linsolve, and a number to be substituted into both the equations and the solutions, and checks if the two calculated values are equal. It works only for the case of a unique solution.

9.12 Checking the Solution

```
def check_solution(equations,solutions):
    eqnlist = equations.split(',')
    if solutions == []:
        print('no solution')
        return
    for eqn in eqnlist:                  # run through given equations
        for sol in solutions:       # run through generated solutions
            var,val = sol.split('=')                       # separate
            var = var.replace(' ','')
            val = val.replace(' ','')
            eqn = eqn.replace(var,'('+ val +')')       # insert value
        left_side,right_side = eqn.split('=')
        difference = left_side + '-('+ right_side +')'
        if mu.multicalc(difference)[0] != '0':
            print('solution false')       # left_side != right_side
            return
    print('solution correct')
```

Matrix Algebra   10

Under suitable conditions, matrices may be added, subtracted, multiplied, and divided, giving rise to an algebraic system with properties similar to those of ordinary algebra. In this chapter we a develop a module `MatAlg.py` that performs these and other operations. The chapter culminates in a matrix calculator. All matrices have Gaussian rational entries. The module is headed by the import statements

```
-------------------------- MatAlg.py --------------------------
import LinSolve as ls
import PolyAlg as pa
import Number as nm
import Arithmetic as ar
import Tools as tl
import math as ma
---------------------------------------------------------------
```

10.1   Elementary Matrix Operations

In this section we define and implement the basic matrix operations. We begin with the easiest: scalar multiplication.

Matrix Scalar Multiple

The *scalar multiple* zA of a matrix A by a number z is the matrix obtained by multiplying each entry of A by z:

$$A = \begin{bmatrix} a_{11} & a_{12} & \cdots & a_{1n} \\ a_{21} & a_{22} & \cdots & a_{2n} \\ \vdots & \vdots & \ddots & \vdots \\ a_{m1} & a_{m2} & \cdots & a_{mn} \end{bmatrix} \quad zA = \begin{bmatrix} za_{11} & za_{12} & \cdots & za_{1n} \\ za_{21} & za_{22} & \cdots & za_{2n} \\ \vdots & \vdots & \ddots & \vdots \\ za_{m1} & za_{m2} & \cdots & za_{mn} \end{bmatrix}$$

The following function implements the operation. It takes a scalar z and a matrix A in standard double list form and returns the scalar multiple zA. (Output in sample run is placed horizontally to save space.)

```
def scalar_mult(z,A):
    B = []
    for row in A:
        Brow = []                          # list for row scalar multiple
        for entry in row:                  # multiply each entry in row by z
            product = ar.main('('+z+')('+entry+')')[0]
            Brow.append(product)
        B.append(Brow)
    return B
```
--------------------------- Sample Run ---------------------------
```
Input:
z = '2/5'
A = '1,i,-3+7i; 22,-5+i,6+(2/3)i; -7+2i,8,1/3.7'
Atab = tl.string2table(A)
print('          A   ')
tl.format_print(Atab, 2, 'right')
print('\n')
print('                   zA')
tl.format_print(scalar_mult(z,A), 2, 'right')
```
```
Output:
            A                                   zA
    1     i    -3+7i               2/5    (2/5)i   -6/5+(14/5)i
   22  -5+i  6+(2/3)i              44/5  -2+(2/5)i  12/5+(4/15)i
 -7+2i    8    1/3.7          -14/5+(4/5)i  16/5        4/37
```

Matrix Factorization

The reverse of scalar multiplication is called *factoring the matrix*. For example the number 2 is factored from each entry in the first matrix, producing twice the second matrix.

$$\begin{bmatrix} 2 & 4 & 6 \\ 8 & 10 & 12 \\ 14 & 16 & 18 \end{bmatrix} = 2 \begin{bmatrix} 1 & 2 & 3 \\ 4 & 5 & 6 \\ 7 & 8 & 9 \end{bmatrix}$$

10.1 Elementary Matrix Operations

The following function takes a matrix A with Gaussian rational entries and returns a matrix with reduced integer entries together with a factor. It first clears any fractions by multiplying the entries of A by the least common multiple of the entry denominators. Then $f_1 = 1/\text{lcm}$ is one of the factors. The function then divides the resulting entries by their greatest common divisor giving rise to a second factor $f_2 = \gcd$. The product $f_1 f_2$ is the desired factor. In the sample run we have included a check.

```
def factor_matrix(A):
    B = tl.copylist(A)
    Bflat = tl.flatten_double_list(B)
    denoms = pa.get_denoms(Bflat)
    lcm = nm.listlcm(denoms)               # get lcm of denoms
    lcm = str(lcm)
    C = scalar_mult(lcm,B)                 # clear denominators
    g = int(C[0][0])
    for row in C:
        for entry in row:                  # get gcd of entries
            g = ma.gcd(g,int(entry))
    r = '1/' + str(g)
    D = scalar_mult(r,C)                   # divide by gcd
    factor = ar.main(str(g) + '/' + lcm)[0]
    factor = tl.add_parens(factor)
    return factor,D

--------------------------- Sample Run ---------------------------
Input:
A = [['6/7', '30/11', '12/5'], ['24/13', '30/17', '18'],
     ['32/9', '132/7', '24/7']]
print('A')
tl.format_print(A,3, 'right'); print('\n')
factor, B = factor_matrix(A)
print('factorization')
print(factor, 'times')
tl.format_print(B,3,'right'); print('\n')
print('check')
tl.format_print(scalar_mult(factor,B),3, 'right')
------------------------------------------------------------------
Output:
A:
     6/7    30/11    12/5
    24/13   30/17     18
    32/9    132/7    24/7

factorization:
(2/765765) times
     328185   1044225    918918
     706860    675675   6891885
    1361360   7220070   1312740
```

```
check:
       6/7    30/11   12/5
      24/13   30/17    18
      32/9   132/7   24/7
```

Matrix Sum and Difference

The *sum* of two $m \times n$ matrices

$$A = \begin{bmatrix} a_{11} & a_{12} & \cdots & a_{1n} \\ a_{21} & a_{22} & \cdots & a_{2n} \\ \vdots & \vdots & \ddots & \vdots \\ a_{m1} & a_{m2} & \cdots & a_{mn} \end{bmatrix}, \quad B = \begin{bmatrix} b_{11} & b_{12} & \cdots & b_{1n} \\ b_{21} & b_{22} & \cdots & b_{2n} \\ \vdots & \vdots & \ddots & \vdots \\ b_{m1} & b_{m2} & \cdots & b_{mn} \end{bmatrix}$$

is defined as

$$A + B = \begin{bmatrix} a_{11}+b_{11} & a_{12}+b_{12} & \cdots & a_{1n}+b_{1n} \\ a_{21}+b_{21} & a_{22}+b_{22} & \cdots & a_{2n}+b_{1n} \\ \vdots & \vdots & \ddots & \vdots \\ a_{m1}+b_{m1} & a_{m2}+b_{m2} & \cdots & a_{mn}+b_{mn} \end{bmatrix}.$$

Thus to find the sum simply add corresponding entries. The *difference* $A - B$ of A and B is defined analogously, corresponding entries subtracted rather than added. Note that the operations can be carried out only if A and B have the same dimensions.

Here are the implementations. (Output in sample run placed horizontally to save space.)

```
def add_mat(A,B):
    C = []
    for i in range(len(A)):
        Crow = []
        for j in range(len(A[0])):
            sum = ar.main(A[i][j]+'+('+B[i][j]+')')[0]
            Crow.append(sum)
        C.append(Crow)
    return C

def subt_mat(A,B):
    C =  scalar_mult('-1',B)
    if A == '0':
        return C
    return add_mat(A,C)
```

```
-------------------------- Sample Run ---------------------------
Input:
A = tl.string2table('1,2,3;4,5,6')
B = tl.string2table('6,5,4;3,2,1')
```

10.1 Elementary Matrix Operations

```
C = add_mat(A,B)
D = subt_mat(A,B)
print('     A')
tl.format_print(A,2,'right')
print('\n')
print('     B')
tl.format_print(B,2,'right')
print('\n')
print('    A + B')
tl.format_print(C,2,'right')
print('\n')
print('    A - B')
tl.format_print(D,2,'right')
```

```
Output:
     A              B           A + B         A - B
   1  2  3       6  5  4       7  7  7      -5 -3 -1
   4  5  6       3  2  1       7  7  7       1  3  5
```

Matrix Product

Let A be an $m \times p$ matrix and B a $p \times n$ matrix:

$$A = \begin{bmatrix} a_{11} & a_{12} & \cdots & a_{1p} \\ a_{21} & a_{22} & \cdots & a_{2p} \\ \vdots & \vdots & \ddots & \vdots \\ a_{m1} & a_{m2} & \cdots & a_{mp} \end{bmatrix}_{m \times p} \quad B = \begin{bmatrix} b_{11} & b_{12} & \cdots & b_{1n} \\ b_{21} & b_{22} & \cdots & b_{2n} \\ \vdots & \vdots & \ddots & \vdots \\ b_{p1} & b_{p2} & \cdots & b_{pn} \end{bmatrix}_{p \times n}$$

The *product* AB of A and B is the $m \times n$ matrix whose (i, j) entry of C is obtained by taking the ith row of A and the jth column of B and multiplying the corresponding entries together:

$$AB = \begin{bmatrix} c_{11} & c_{12} & \cdots & c_{1n} \\ c_{21} & c_{22} & \cdots & c_{2n} \\ \vdots & \vdots & \ddots & \vdots \\ c_{m1} & c_{m2} & \cdots & c_{mn} \end{bmatrix}_{m \times n} \quad c_{ij} := a_{i1}b_{1j} + a_{i2}b_{2j} + a_{ip}b_{pj}.$$

We have attached the matrix dimensions to highlight their relationship.

Here is the code for multiplying matrices together with a sample run (output of sample run displayed side by side). For greater flexibility the function allows for matrix scalar multiplication as well.

```
def mult_mat(A,B):
    if isinstance(A,str):
        return scalar_mult(A,B)
```

```
            if isinstance(B,str):
                return scalar_mult(B,A)
            C = []
            for i in range(len(A)):                    # run through rows of A
                Crow = []
                for j in range(len(B[0])):             # run through columns of B
                    s = '0'
                    for k in range(len(A[0])):
                        s = s + '+('+A[i][k]+')('+B[k][j]+')'
                    s = ar.main(s)[0]
                    Crow.append(s)
                C.append(Crow)
            return C

-------------------------- Sample Run --------------------------
Input:
A = tl.string2table('1,2,3,4;5,6,7,8;9,10,11,12')
I = tl.string2table('1,0,0,0;0,1,0,0;0,0,1,0;0,0,0,1')
AI = mult_mat(A,I)
print('         A')
tl.format_print(A,2,'right');print('\n')
print('         I')
tl.format_print(I,2,'right');print('\n')
print('         AI')
tl.format_print(AI,2,'right'); print('\n')
I = tl.string2table('1,0,0;0,1,0,;0,0,1')
IA = mult_mat(I,A)
tl.format_print(I,2,'right');print('\n')
print('         IA')
tl.format_print(IA,2,'right');print('\n')
----------------------------------------------------------------
Output:
       A                    I               AI
  1   2   3   4      1  0  0  0       1   2   3   4
  5   6   7   8      0  1  0  0       5   6   7   8
  9  10  11  12      0  0  1  0       9  10  11  12
                     0  0  0  1

     I                  IA
  1  0  0          1   2   3   4
  0  1  0          5   6   7   8
  0  0  1          9  10  11  12
----------------------------------------------------------------
```

The matrices I in the sample run are called *identity matrices of orders 4 and 3*, respectively. In general, the $n \times n$ *identity matrix* I has 1's down the main diagonal and 0's elsewhere. It is also denoted by I_n, the subscript indicating its order. As the sample run suggests, $AI_n = I_m A = A$ for any $m \times n$ matrix A. The identity matrix is the analog of the number 1 in ordinary algebra.

10.2 The Inverse of a Matrix

The *inverse* of an $n \times n$ matrix A is an $n \times n$ matrix X with the property that $AX = I$, where I is the $n \times n$ identity matrix. The inverse is the analog of the reciprocal of a nonzero number in ordinary algebra. If the inverse exists then it is unique and the equation $XA = I$ also holds.

To find an algorithm for the construction of X we consider first a 2×2 matrix A and write the equation $AX = I$ explicitly as

$$\begin{bmatrix} a & b \\ c & d \end{bmatrix} \begin{bmatrix} x_1 & x_2 \\ x_3 & x_4 \end{bmatrix} = \begin{bmatrix} 1 & 0 \\ 0 & 1 \end{bmatrix}.$$

The entries x_i are the unknown entries of the inverse X. Multiplying and equating entries yields a system in the variables x_1 and x_3 and another in x_2 and x_4:

$$ax_1 + bx_3 = 1 \qquad ax_2 + bx_4 = 0$$
$$cx_1 + dx_3 = 0 \quad \text{and} \quad cx_2 + dx_4 = 1$$

These can be solved by forming the augmented matrix of each system,

$$\begin{bmatrix} a & b & 1 \\ c & d & 0 \end{bmatrix} \quad \text{and} \quad \begin{bmatrix} a & b & 0 \\ c & d & 1 \end{bmatrix},$$

and then applying the row echelon algorithm twice yielding

$$\begin{bmatrix} 1 & 0 & x_1 \\ 0 & 1 & x_3 \end{bmatrix} \quad \text{and} \quad \begin{bmatrix} 1 & 0 & x_2 \\ 0 & 1 & x_4 \end{bmatrix},$$

the solutions appearing in the last column. This is equivalent to applying the row echelon algorithm to the combined matrix

$$[A \mid I] = \begin{bmatrix} a & b & 1 & 0 \\ c & d & 0 & 1 \end{bmatrix}.$$

If A has an inverse, then the algorithm yields the solution of both systems:

$$\begin{bmatrix} 1 & 0 & x_1 & x_2 \\ 0 & 1 & x_3 & x_4 \end{bmatrix}.$$

The inverse matrix X appears on the right and the identity matrix on the left. The process works for an arbitrary $n \times n$ matrix A: Reduce the combined augmented matrix $[A|I]$ to echelon form, yielding a matrix $[B|X]$. If B is the identity matrix, then X is the inverse of A; otherwise, A has no inverse. The standard notation for the inverse of A is A^{-1}.

The function invert_mat implements the algorithm. It takes a square matrix A, attaches the identity matrix on the right, and then invokes row_echelon. If the left half is the identity matrix, then the right half is the inverse.

```
def invert_mat(A):
    nrows = len(A)
    B = attach_id(A)              # attach identity I to the right of A
    C = ls.row_echelon(B)[0]                             # row reduce
    left,right = split_mat(C)     # extract left and right parts
    I = makeid(nrows)
    if left == I:                 # compare left part with I
        return right                              # inverse exists
    else: return []                           # inverse fails to exist

--------------------------- Sample Run ---------------------------
Input:
A = '1+2i,2,3;4,5+3i,6;7,8,9'
print('A')
A = tl.string2table(A)
tl.format_print(A,4,'left')
print('\n')
X = invert_mat(A)
print('A^(-1)')
tl.format_print(X,4,'left')
print('\n')
print('check AA^(-1):')
tl.format_print(mult_mat(A,X),4,'left')
```

Output:
A
1+2i   2   3
4   5+3i   6
7   8   9+4i

A^(-1)
-519/4330-(769/2165)i   -3/4330+(329/4330)i   51/866+(18/433)i
153/4330+(541/4330)i   101/4330-(973/4330)i   15/866+(87/866)i
489/4330+(249/2165)i   153/4330+(541/4330)i   -3/866-(52/433)i

check AA^(-1):
1   0   0
0   1   0
0   0   1

10.2 The Inverse of a Matrix

The following function returns the identity matrix for any specified dimension n.

```
def makeid(n):
    global I
    I = [['0' for j in range(n)] for i in range(n)]    # zero matrix
    for i in range(n):
        for j in range(n):
            if i == j: I[i][j] = '1'  # 1's along the diagonal
    return I
```

The function `attach_id(A)` attaches the identity matrix to the right of A. The process is simplified by attaching the identity to the bottom of the transpose of A and then taking the transpose of the result.

```
def attach_id(A):
    nrows =len(A)
    I = makeid(nrows)
    B = ls.transpose(A)
    C = []
    for i in range(nrows):           # put rows of B in C
        C.append(B[i])
    for row in I:                    # put I below B
        C.append(row)
    return ls.transpose(C)
```

The function `split_mat(A,col)` returns the right and left parts of matrix A separated at column col, that column placed in the left part. (Output in sample run placed horizontally to save space.)

```
def split_mat(A):                    # split into right and left parts
    B = transpose(A)
    nrows = len(A)
    top = []; bottom = []
    for i in range(nrows):
        top.append(B[i])
    for i in range(nrows,2*nrows):
        bottom.append(B[i])
    return ls.transpose(top), ls.transpose(bottom)
```
---------------------------- Sample Run ----------------------------
```
Input:
A = tl.string2table('a,b,c,d,e,f; g,h,i,j,k,l;m,n,o,p,q,r')
left_part,right_part = split_mat(A)
tl.format_print(A,1,'left'); print('\n')
```

```
tl.format_print(left_part,1,'left'); print('\n')
tl.format_print(right_part,1,'left')
```
```
Output:
a b c d e f       a b c   d e f
g h i j k l       g h i   j k l
m n o p q r       m n o   p q r
```

10.3 Matrix Exponentiation

The following function returns positive or negative powers p of a square matrix A. It handles the case $p < 0$ by using the easily established fact that $A^p = (A^{-1})^q$ where $q = -p > 0$. The sample run illustrates the general rule $A^p A^q = A^{p+q}$ with the example $A^7 A^{-3} = A^4$.

```
def power_mat(A,p):
    if p == 0:
        return makeid(len(A))
    if p == 1:
        return A
    B = tl.copylist(A)                          # don't alter A
    if p < 0:
        B = invert_mat(A);
        p = -p
    if B == []: return []
    prod = B
    for i in range(p-1):                        # multiply B times itself p-1 times
        prod = mult_mat(prod,B)
    return prod
```
```
--------------------------- Sample Run ---------------------------
Input:
A = tl.string2table('3,1,2;5,4,7;-1,8,3')
B = power_mat(A,7)
C = power_mat(A,-3)
D = mult_mat(B,C)
E = power_mat(A,4)
print('A')
tl.format_print(A,2,'right'); print('\n')
print('A^7')
tl.format_print(B,2,'right'); print('\n')
print('A^(-3)')
tl.format_print(C,2,'right'); print('\n')
print('A^7*A^(-3)')
tl.format_print(D,2,'right'); print('\n')
print('A^4')
tl.format_print(E,2,'right')
```

```
Output:
3  1  2
5  4  7
-1 8  3

A^7
2330556  4974196   4520096
7379804  15748768  14312476
6485540  13858376  12584004

A^(-3)
755/3267    -4771/71874  -262/35937
43/594      -199/6534    61/6534
-1297/6534  2459/35937   -212/35937

A^7*A^(-3)
1434  3092  2782
4498  9818  8792
4090  8482  7860

A^4
1434  3092  2782
4498  9818  8792
4090  8482  7860
```

10.4 Solving Systems Using Matrix Inversion

The inverse matrix operation may be used to solve systems of linear equations that have the same number of unknowns as equations. First write the system in (9.2) as a matrix equation $AX = B$, where

$$A = \begin{bmatrix} a_{11} & a_{12} & \cdots & a_{1n} \\ a_{21} & a_{22} & \cdots & a_{2n} \\ \vdots & \vdots & \ddots & \vdots \\ a_{m1} & a_{m2} & \cdots & a_{mn} \end{bmatrix}, \quad X = \begin{bmatrix} x_1 \\ x_2 \\ \vdots \\ x_n \end{bmatrix}, \quad \text{and } B = \begin{bmatrix} b_1 \\ b_2 \\ \vdots \\ b_m \end{bmatrix}.$$

The matrix A is called the *coefficient matrix* of the system. If $m = n$ and A^{-1} exists, then, as in ordinary algebra, one can multiply the equation $AX = B$ on the left by A^{-1}, resulting in $A^{-1}AX = A^{-1}B$. Since $A^{-1}AX = IX = X$, the solution of the system is given by the matrix equation $X = A^{-1}B$. This idea is sometimes useful in matrix calculations.

10.5 A Matrix Calculator

The functions in this section evaluate algebraic expressions involving matrices. In the example

$$(3+1.2i)^{\wedge}(-3)(A^{\wedge}2C - 2D3E^{\wedge}(-5))F + (1-i)A^{\wedge}0,$$

the capital letters are symbols for matrices entered by the user. The symbol A^0 (for a square matrix A) denotes the identity matrix with dimensions those of A. The dimensions of a matrix can be quite large, limited only by computer memory and speed. The user enters matrices in a list, as shown in the sample run. The list is converted into a dictionary by the function load_dict. The function print_mat_dict() is included for illustration. (Output matrices are displayed side by side to save space.)

```
def load_dict(matrices):
    global mat_dict
    mat_dict = {}
    for mat in matrices:                 # run through set of matrices
        mat = mat.replace(' ','')                 # remove white space
        label,entries = mat.split('=')   # separate matrix and label
        mat_dict[label] = tl.string2table(entries)   # enter in dict.

def print_mat_dict():           # run through dictionary, print items
    global mat_dict
    for item in mat_dict:
        print('    ',item)         # print the letter, then the matrix
        tl.format_print(mat_dict[item],2,'right')
        print('\n')
```

--------------------------- Sample Run ---------------------------
Input:
matrices = ['A = 1,0,3;0,5,6;7,8,0',\
            'B = 0,2,3;4,6,0;7,0,9', \
            'C = -5,0,7;8,-1,0;4,3,0']
load_dict(matrices)
print_mat_dict()

Output:
```
     A              B              C
  1  0  3        0  2  3       -5  0  7
  0  5  6        4  6  0        8 -1  0
  7  8  0        7  0  9        4  3  0
```

10.5 A Matrix Calculator

Here is the main function for the calculator. The sample runs illustrate various general laws of matrix algebra. For example, the law $(A + B)^t = A^t + B^t$ is illustrated in the sample run by showing that $(A + B)^t - A^t - B^t$ is the zero matrix.

```
def calculator(expression,matrices):
    global expr, idx, mat_dict
    load_dict(matrices)
    expr = expression
    expr = expr.replace(' ','')
    expr = tl.attach_missing_exp(expr,tl.upper)
    expr = tl.fix_signs(expr)
    expr = tl.fix_operands(expr)
    expr = tl.insert_asterisks(expr,tl.upper)
    idx = 0
    r = allocate_ops(0)
    return r
```

```
--------------------------- Sample Run ---------------------------
Input:
d1 = '(A+B)^t- A^t - B^t'            # (A+B)^t = A^t+B^t
d2 = '(AB)^t- (B^t)A^t'              # (AB)^t = (B^t)(A^t)
d3 = '(AB)^(-1)-B^(-1)A^(-1)'        # (AB)^(-1)= B^(-1)A^(-1)
d4 = '(A^t)^(-1) - (A^(-1))^t'       # (A^t)^(-1)=(A^(-1))^t
d5 = '(A+B)C - AC-BC'                # (A+B)C=AC+BC
d6 = '(A^t)^(-1) - (A^(-1))^t'       # (A^t)^(-1)=(A^(-1))^t
D = [d1,d2,d3,d4,d5,d6]
matrices = ['A = 3,0,1  ;7,6,0  ;0,8,9',\
            'B = 0,2,3  ;4,0,6  ;7,8,0',\
            'C = 2,0,-3 ;0,6,-2 ;-7,4,0']
for d in D:
    tl.format_print(calculator(d,matrices),2,'right')
    print('\n')
------------------------------------------------------------------
Output:                              # same for each d
0 0 0
0 0 0
0 0 0
------------------------------------------------------------------
```

Here is the function that assigns the calculations. It is similar in concept to the corresponding functions in Arithmetic.py, PolyAlg.py and MultiAlg.py.

```
def allocate_ops(mode):
    global expr, idx, mat_dict
    r = []
    while idx < len(expr):
        ch = expr[idx]
        if tl.isnumeric(ch):
            z,idx = tl.extract_numeric(expr,idx)
```

```
                    r = ar.maincalc(z)[0]
              elif tl.isupper(ch):
                    r = mat_dict[ch]
                    idx += 1
              elif ch == '+':
                    if mode > 0: break               # wait for higher mode
                    idx += 1
                    s = allocate_ops(0)
                    r = add_mat(r,s)
              elif ch == '-':
                    if mode > 0: break               # wait for higher mode
                    idx += 1
                    s = allocate_ops(1)
                    r = subt_mat(r,s)
              elif ch == '*':
                    idx += 1
                    s = allocate_ops(1)
                    r = mult_mat(r,s)
              elif ch == '^':                        # highest mode
                    if expr[idx+1] == 't':
                         r = ls.transpose(r)
                         idx+=2                      # skip '^t' and move on
                         continue
                    exp,idx = tl.extract_exp(expr,idx)
                    exp = ar.main(exp)[0]            # fix exp = 0-1
                    if isinstance(r,str):
                         r = ar.main( '('+ r + ')^(' + exp + ')')[0]
                    else:
                         r = power_mat(r, int(exp))
                         if r == []: return r        # inverse may not exist
              elif ch == '(':
                    start = idx
                    paren_expr,end = tl.extract_paren(expr,start)
                    if tl.isarithmetic(paren_expr):
                         r = ar.main(paren_expr)[0]
                         idx = end
                    else:
                         idx+=1
                         r = allocate_ops(0)
                         idx+=1
              elif ch == ')': break
        return r
```

10.6 Application: Moore-Penrose Inverse

An $m \times n$ system of equations $AX = B$ may not have a solution, but it is possible to find a matrix X such that a common error measurement $\|AX - B\|$ is as small as possible. The matrix X is then called a *least squares solution* of the equation. The quantity $\|AX - B\|$ is

10.6 Application: Moore-Penrose Inverse

the square root of sum of the squares of the entries of $AX - B$ and is called the *Euclidean distance* from AX to B. It may be shown that a least squares solution is given by the product $X = A^+ B$, where

$$A^+ = D^t(DD^t)^{-1}(C^tC)^{-1}C^t,$$

where C and D are described below. Additionally, for this solution, $\|X\|$ is as small as possible, giving us the unique *minimal least squares solution* of $AX = B$. The matrix A^+ is called the *Moore-Penrose inverse of A*.

To describe C and D, let R be the reduced row echelon form of A and let $j_1 < j_2 < \ldots < j_r$ denote the columns in which a leading entry appears. It may be shown that $A = CD$, where C is the $m \times r$ matrix obtained from A by deleting all but the columns $j_1, j_2, \ldots j_r$, and D is the $r \times n$ matrix obtained from R by deleting the last $m - r$ rows (the zero rows). For example, for the matrix

$$A = \begin{bmatrix} 1 & 2 & 3 & 12 \\ 4 & 5 & 6 & 15 \\ 7 & 8 & 9 & 18 \end{bmatrix}$$

we have

$$R = \begin{bmatrix} 1 & 0 & -1 & -10 \\ 0 & 1 & 2 & 11 \\ 0 & 0 & 0 & 0 \end{bmatrix},$$

hence $r = 2$, $j_1 = 1$, and $j_2 = 2$, and the matrices C and D are

$$C = \begin{bmatrix} 1 & 2 \\ 4 & 5 \\ 7 & 8 \end{bmatrix} \quad \text{and} \quad D = \begin{bmatrix} 1 & 0 & -1 & -10 \\ 0 & 1 & 2 & 11 \end{bmatrix}.$$

The reader may check that $A = CD$.

The function decompose(A) prints out C and D for any matrix A.

```
def decompose(A):
    R = ls.row_echelon(A)[0]
    lead_cols = leading_entry_cols(R)    # leading entry col numbers
    r = len(lead_cols);  m = len(A)
    C = delete_cols(A, lead_cols)        # delete all but these cols of A
    D = delete_last_rows(R,m-r)          # delete rows r+1,...,m
    return C,D,R

def delete_cols(A,col_list):
    ncols = len(A[0])
    At = transpose(A)
    B = []
    for j in range(ncols):
        if j in col_list:
            B.append(At[j])
    C = transpose(B)
```

```
        return C

def delete_last_rows(R,k):
    D = []
    for i in range(k+1):
        D.append(R[i])
    return D

def leading_entry_cols(R):
    lead_cols = []
    for i in range(len(R)):
        for j in range(len(R)):
            if R[i][j] != '0':
                lead_cols.append(j)
                break
    return lead_cols
```

The function moore_penrose(A) returns A^+.

```
def moore_penrose(A):
    C,D,R = decompose(A)
    C = tl.table2string(C)
    D = tl.table2string(D)
    Aplus = 'D^t(DD^t)^(-1)(C^tC)^(-1)C^t'
    matrices = ['C='+C,'D='+D]
    return calculator(Aplus, matrices)
```

--------------------------- Sample Run ---------------------------
```
Input:
A = tl.string2table('4,5,6;1,2,3;7,8,9')
B = tl.string2table('12,9,14')
print('A')
tl.format_print(A,3, 'right'); print('\n')
print('B^t')
tl.format_print(transpose(B),3, 'right'); print('\n')
Aplus = moore_penrose(A)
print('          Aplus')
tl.format_print(Aplus,3,'right'); print('\n')
LS = mult_mat(Aplus,transpose(B))
print('   minimal least squares solution')
tl.format_print(LS,3,'right')
```

```
Output:
A                B^t         Aplus
4   5   6        12       -1/6   -23/36   11/36
1   2   3         9          0    -1/18    1/18
7   8   9        14        1/6    19/36   -7/36
```

```
minimal least squares solution:
-125/36
    5/18
 145/36
```

10.7 Application: Curve Fitting

Closely related to the notion of interpolation (see Sect. 6.9) is the problem of finding a function f from a prescribed class, typically polynomials, whose deviation from a given set data pairs (x_k, y_k) is as small as possible. Here, contrary to the case of a interpolation function, it may not be true that $f(x_k) = y_k$ for all k. Instead, f minimizes a certain error measurement. The most common choice of this measure, one that readily lends itself to mathematical analysis, is the square of Euclidean distance introduced in the preceding section:

$$\sum_{k=1}^{n} |f(x_k) - y_k|^2, \quad n = \text{number of data points}(x_k, y_k). \tag{10.1}$$

Using this measure results in the *least squares method of curve fitting*. In this section we consider the problem of fitting data to polynomials and construct a function `poly_fit(data,m)` that generates, for each $m < n$, the polynomial of degree m that best fits the data in the least squares sense.

Consider first the problem of finding a cubic polynomial[1]

$$P(x) = c_0 + c_1 x + c_2 x^2 + c_3 x^3$$

that best fits the data

$$(x_1, y_1), (x_2, y_2), \ldots, (x_n, y_n), \text{ where } x_j \neq x_k \text{ for } j \neq k.$$

We seek coefficients c_0, c_1, c_2, c_3 that minimize

$$\sum_{i=1}^{n} \left[P(x_i) - y_i\right]^2 = \sum_{i=1}^{n} \left[c_0 + c_1 x_i + c_2 x_i^2 + c_3 x_i^3 - y_i\right]^2. \tag{10.2}$$

Using calculus one may show that the coefficients c_i are the solutions of the linear system

---

[1] We have written $P(x)$ with leading term last to simplify notation in subsequent expressions.

$$\sum_{i=1}^{n}\left[c_0 + c_1 x_i + c_2 x_i^2 + c_3 x_i^3 - y_i\right] = 0$$

$$\sum_{i=1}^{n} x_i\left[c_0 + c_1 x_i + c_2 x_i^2 + c_3 x_i^3 - y_i\right] = 0$$

$$\sum_{i=1}^{n} x_i^2\left[c_0 + c_1 x_i + c_2 x_i^2 + c_3 x_i^3 - y_i\right] = 0$$

$$\sum_{i=1}^{n} x_i^3\left[c_0 + c_1 x_i + c_2 x_i^2 + c_3 x_i^3 - y_i\right] = 0.$$

Setting

$$S_k = \sum_{i=1}^{n} x_i^k \text{ and } T_k = \sum_{i=1}^{n} x_i^k y_i. \tag{10.3}$$

we can write the system as

$$c_0 S_0 + c_1 S_1 + c_2 S_2 + c_3 S_3 = T_0$$
$$c_0 S_1 + c_1 S_2 + c_2 S_3 + c_3 S_4 = T_1$$
$$c_0 S_2 + c_1 S_3 + c_2 S_4 + c_3 S_5 = T_2$$
$$c_0 S_3 + c_1 S_4 + c_2 S_5 + c_3 S_6 = T_3$$

The same analysis works for polynomials of degree m, resulting in the system

$$c_0 S_0 + c_1 S_1 + \cdots + c_m S_m = T_0$$
$$c_0 S_1 + c_1 S_2 + \cdots + c_m S_{m+1} = T_1$$
$$\vdots$$
$$c_0 S_m + c_1 S_{m+1} + \cdots + c_m S_{2m} = T_m$$
(10.4)

This may be written as a matrix equation $SC = T$, where

$$S = \begin{bmatrix} S_0 & S_1 & \cdots & S_m \\ S_1 & S_2 & \cdots & S_{m+1} \\ \vdots & \vdots & \cdots & \vdots \\ S_m & S_{m+1} & \cdots & S_{2m} \end{bmatrix}, \quad C = \begin{bmatrix} c_0 \\ c_1 \\ \vdots \\ c_m \end{bmatrix} \text{ and } T = \begin{bmatrix} T_0 \\ T_1 \\ \vdots \\ T_m \end{bmatrix}$$

The polynomial of best fit has coefficients given by the matrix $C = S^{-1} T$.

For the matrix S to be invertible it is necessary that the degree m of the desired polynomial be less than the number n of data points. If $m = n - 1$, then one can choose the coefficients to obtain an *exact* fit: $P(x_k) = y_k$ for each i. In this case the error term

10.7 Application: Curve Fitting

$$\sum_{i=1}^{n} [P(x_i) - y_i]^2 = \sum_{i=1}^{n} [c_0 + c_1 x_i + c_2 x_i^2 \cdots + c_{n-1} x_i^{n-1} - y_i]^2$$

is zero, which implies that each term in the sum is zero. This condition gives rise to the system

$$\begin{bmatrix} 1 & x_1 & x_1^2 & \cdots & x_1^{n-1} \\ 1 & x_2 & x_2^2 & \cdots & x_2^{n-1} \\ & & \vdots & & \\ 1 & x_n & x_n^2 & \cdots & x_n^{n-1} \end{bmatrix} \begin{bmatrix} c_0 \\ c_1 \\ \vdots \\ c_{n-1} \end{bmatrix} = \begin{bmatrix} y_1 \\ y_2 \\ \vdots \\ y_n \end{bmatrix}. \quad (10.5)$$

Thus $C = V^{-1} Y$, where V is the $n \times n$ matrix on the left in (10.5), called a *Vandermonde matrix*. The inverse exists because the x_i's are distinct. Indeed, using row properties of determinants (developed in the next chapter) one easily shows that the determinant of V, the so-called *Vandermonde determinant*, is

$$\prod_{1 \leq i < j \leq n} (x_j - x_i),$$

which is not zero since $x_j - x_i$ for $i \neq j$.

The function poly_fit(data,m) takes data in the form of a string of ordered pairs and returns the polynomial of degree m of best fit. The sample run of the function produces best fit polynomials of various degrees for a fixed set of data points. As mentioned earlier, if $m = n - 1$ one obtains a polynomial that gives an exact fit. In fact, in this case one obtains the Lagrange interpolating polynomial, as the sample run illustrates.

```
def poly_fit(data,m):
    # requires m < len(data)
    # get exact fit when m = len(data)-1
    data = data2list(data)
    if len(data) <= m:
        return
    s_mat = S_mat(m,data)
    t_mat = T_mat(m,data)
    s_mat_inv = invert_mat(s_mat)
    C = mult_mat(s_mat_inv,t_mat)            # C = s_mat^(-1)*t_mat
    flist = tl.flatten_double_list(C)
    flist = flist[::-1]                      # reverse flist
    pa.var = 'x'
    return pa.flist2pol(flist)               # convert list to pol

------------------------- Sample Run -------------------------
Input:
data = '(1,3),(2,1.1),(3,5),(4,4.2),(5,9),(6,8.1)'
print('degree 1  ',poly_fit(data,1),'\n')
print('degree 2  ',poly_fit(data,2),'\n')
print('degree 3  ',poly_fit(data,3),'\n')
```

```
print('degree 4  ',poly_fit(data,4),'\n')
print('degree 5  ',poly_fit(data,5),'\n')
print('lagrange  ',pa.lagrange_interp(data),'\n')
print('degree 6  ',poly_fit(data,6))
```

```
Output:
degree 1    (242/175)x+17/75

degree 2    (43/280)x^2+(431/1400)x+83/50

degree 3    (-133/540)x^3+(863/315)x^2+(-28349/3780)x+118/15

degree 4    (-1/60)x^4+(-7/540)x^3+(293/180)x^2+(-20537/3780)x+20/3

degree 5    (-53/150)x^5+(37/6)x^4+(-809/20)x^3+(3697/30)x^2+ \
            (-50999/300)x+422/5

lagrange    (-53/150)x^5+(37/6)x^4+(-809/20)x^3+(3697/30)x^2+ \
            (-50999/300)x+422/5
none
```

The function comparisons(pols,data,p) takes a list pols of polynomials P generated by poly_fit(data,m) and prints for each value of $m < n$ a table that displays the data values (x_k, y_k) and the polynomial values

$$z_k = P(x_k), \quad k = 1, 2, \ldots, n, \quad P \text{ in pols}.$$

It also calculates least squares error in using z_k to approximate x_k. Decimals with p places, rather than fractions, are used for ease of comparison. The sample run uses the polynomials and data from the previous sample run.

```
def comparisons(pols,data,p):
    data = pa.data2lists(data)
    for m in range(1,len(pols)):     # run through the list of pols
        table = []
        pol = pols[m]
        error = '0'                                  # initialize
        for k in range(len(data)):   # run through the data points
            xk = data[k][0]; yk = data[k][1]
            xk = ar.main(xk)[0]      # convert x,y data to fracs
            yk = ar.main(yk)[0]
            zk = ar.evaluate(pol,xk,p)[0]   # value of pol at x data
            diff_squared = '((' + zk + ')-(' + yk + '))^2'
            error = ar.main(error + '+' + diff_squared)[0]
            yk = ar.decimal_approx(yk,p)[0]          # approx. y data
            zk = ar.decimal_approx(zk,p)[0]
            table.append(['  ',xk,yk,zk, '  '])
```

10.7 Application: Curve Fitting

```
            error = ar.decimal_approx(error,p)[0]
            header = ['m = '+str(m), 'xk','yk','zk','error = '+ error]
            table = [header] + table
            tl.format_print(table, 4, 'right'); print('\n')
```

```
-------------------------- Sample Run --------------------------
Input:
pols = [pol1,pol2,pol3,pol4,pol5]       # generated in previous run
comparisons(pols,data,4)                # four decimal places
----------------------------------------------------------------
Output:
m = 1      xk      yk         zk     error = 11.9681
            1       3      1.6095
            2     1.1      2.9923
            3       5      4.3752
            4     4.2       5.758
            5       9      7.1409
            6     8.1      8.5238

m = 2      xk      yk         zk     error = 11.0877
            1       3      2.1214
            2     1.1        2.89
            3       5      3.9657
            4     4.2      5.3485
            5       9      7.0385
            6     8.1      9.0357

m = 3      xk      yk         zk     error = 7.1568
            1       3      2.8603
            2     1.1      1.8555
            3       5      3.3746
            4     4.2      5.9396
            5       9       8.073
            6     8.1      8.2968

m = 4      xk      yk         zk     error = 7.1339
            1       3      2.8317
            2     1.1      1.9412
            3       5      3.3174
            4     4.2      5.8825
            5       9      8.1587
            6     8.1      8.2682

m = 5      xk      yk       zk       error = 0
            1       3        3
            2     1.1      1.1
            3       5        5
            4     4.2      4.2
            5       9        9
            6     8.1      8.1
----------------------------------------------------------------
```

For the remainder of the section we construct the supporting functions required by poly_fit. The functions S and T calculate the sums S_k and T_k. Data is entered as in the module lagrange_interp.py and converted by data2lists.

```
def S(k,data):                                       # returns S_k
    s_sum = '0'
    for j in range(len(data)):
        x = data[j][0]
        power = x + '^' + str(k)
        s_sum = ar.main(s_sum + '+' + power)[0]
    return s_sum

def T(k,data):                                       # returns T_k
    t_sum = '0'
    for j in range(len(data)):
        x = data[j][0]; y = data[j][1]
        power_prod = '('+ x + '^'+ str(k)+ ')*('+ y + ')'
        t_sum = ar.main(t_sum + '+' + power_prod)[0]
    return t_sum
```

The following functions return the matrices S and T, respectively.

```
def S_mat(m,data):               # returns matrix of S_k entries
    s_mat = []
    for i in range(m+1):
        row = []
        for j in range(m+1):
            row.append(S(i+j,data))
        s_mat.append(row)
    return s_mat

def T_mat(m,data):           # returns column matrix of T_k entries
    t_mat = []
    for k in range(m+1):
        t_mat.append([T(k,data)])
    return t_mat
```

10.8 Elementary Matrices

An *elementary matrix* is one that can be gotten from the identity matrix by a row operation. Their importance derives from the fact that a row operation on a matrix A can be achieved by multiplying A on the left by the elementary matrix produced by performing the operation on

10.8 Elementary Matrices

1. The following example illustrates this. The first three matrices are the result of applying row operations to the identity I:

$$I_{r1<->r3} = \begin{bmatrix} 0 & 0 & 1 \\ 0 & 1 & 0 \\ 1 & 0 & 0 \end{bmatrix}, I_{(5)r2} = \begin{bmatrix} 1 & 0 & 0 \\ 0 & 5 & 0 \\ 0 & 0 & 1 \end{bmatrix}, I_{(7)r1+r2} = \begin{bmatrix} 1 & 0 & 0 \\ 7 & 1 & 0 \\ 0 & 0 & 1 \end{bmatrix}.$$

If we let

$$A = \begin{bmatrix} 1 & 2 \\ 3 & 4 \\ 5 & 6 \end{bmatrix}$$

then we can produce same row operations on A by multiplying A on the left by the above elementary matrices:

$$I_{r1<->r3} A = \begin{bmatrix} 5 & 6 \\ 1 & 2 \\ 3 & 4 \end{bmatrix} = A_{r1<->r3}, \quad I_{(5)r2} A = \begin{bmatrix} 1 & 2 \\ 15 & 20 \\ 5 & 6 \end{bmatrix} = A_{(5)r2},$$

$$I_{(7)r1+r2} A = \begin{bmatrix} 1 & 2 \\ 10 & 18 \\ 5 & 6 \end{bmatrix} = A_{(7)r1+r2}.$$

In general, for any matrix A, there exists a sequence $E_1, E_2, \ldots, E_n$ of elementary matrices such that the product $E_1 E_2 \cdots E_n A$ is the reduced row echelon form of A.

The following function produces elementary matrices using the ops output of row_echelon(A). The sample run illustrates the aforementioned property.

```
def elementary_matrices(A):
    # returns list of elementary matrices and their product
    nrows = len(A)
    I = makeid(nrows)           # ops are applied to identity matrix
    B, ops = ls.row_echelon(A)  # echelon matrix and operations
    E = []                                   # list of elementary
    Eprod = I                                # initialize
    n = 1                                    # number for label
    for op in ops:
        C = ls.row_op_calc(op,I)             # perform op on I
        print('     E'+ str(n)); n+=1        # label
        tl.format_print(C,2,'right')  # print elementary matrix C
        print('\n')
        E.append(C)                          # keep C
        Eprod = mult_mat(C,Eprod)            # update product
    return E, Eprod       # elementary matrices and their product
```

```
-------------------------- Sample Run --------------------------
Input:
A = tl.string2table('-2,0,3;0,5,6;7,8,0')
E,Eprod = elementary_matrices(A)                  # run the program
print('         Eprod')
tl.format_print(Eprod,2,'right')
print('\n')
print('   Eprod*A')                      # Eprod*A is the echelon form I
tl.format_print(mult_mat(Eprod,A),2,'right')
----------------------------------------------------------------
Output:
      E1              E2              E3              E4              E5
  -1/2  0  0        1  0  0        1    0  0        1   0  0        1  0    0
     0  1  0        0  1  0        0  1/5  0        0   1  0        0  1    0
     0  0  1       -7  0  1        0    0  1        0  -8  1        0  0  10/9

      E6              E7                 Eprod              Eprod*A
   1  0  3/2       1  0    0       16/3   -8/3   5/3       1  0  0
   0  1  0         0  1  -6/5     -14/3    7/3  -4/3       0  1  0
   0  0  1         0  0    1       35/9  -16/9  10/9       0  0  1
----------------------------------------------------------------
```

Vectors 11

An ordered list $v = [a_1, a_2, \ldots, a_n]$ of numbers (also called *scalars* in the present context), is called an *n-dimensional vector*. The numbers a_k are called the *components* of the vector. We identify this mathematical list with the corresponding Python list. In particular, we may write the components of v as v[0],v[1],...,v[n]. Two n-dimensional vectors u,v are said to be *equal* if u[k] = v[k] for all k. The vector whose components are all zero is called the *zero vector*. For purposes of coding, components of vectors are taken to be Gaussian rationals.

In this chapter we describe some basic properties of vectors and develop functions that implement some of associated computational processes. The chapter focusses on the notion of linear independence of matrix rows and on the range and kernel of a matrix. The functions in the chapter comprise the module Vectors.py, headed by

```
------------------------- Vectors.py -------------------------
import LinSolve as ls
import Arithmetic as ar
import MultiAlg as mu
import MatAlg as mat
import Tools as tl
--------------------------------------------------------------
```

11.1 Linear Combinations

Vectors may be added, subtracted, and multiplied by a scalar in the same way as matrices, namely termwise:

$$z[a_1, a_2, \ldots, a_n] = [za_1, za_2, \ldots, za_n]$$
$$[a_1, a_2, \ldots, a_n] \pm [b_1, b_2, \ldots, b_n] = [a_1 \pm b_1, a_2 \pm b_2, \ldots, a_n \pm b_n].$$

Sums of more than two vectors are defined analogously. Here are functions that implement the operations.

```
def scalar_mult_vec(z,v):
    u = []
    for i in range(len(v)):
        prod = '(' + z + ')('+ v[i] + ')'
        u.append(ar.main(prod)[0])
    return u

def add_vec(u,v):
    vecsum = []
    for i in range(len(u)):
        s = u[i] + '+' + '(' + v[i] + ')'
        vecsum.append(ar.main(s)[0] )
    return vecsum

def add_vecs(vecs):               # add a list vecs of vectors
    vecsum = vecs[0]
    for i in range(1,len(vecs)):
        vecsum = add_vec(vecsum,vecs[i])
    return vecsum

def subt_vec(u,v):
    w = scalar_mult_vec('(-1)',v)
    return add_vec(u,w)
```

A *linear combination* of vectors is a sum of *scalar multiples* of the vectors. For example, the vector

$$(5.7)[1, 2, 3] - [4, 5, 6] + (3/2 + i)[7, 8, 9]$$

is a linear combination of $[1, 2, 3]$, $[4, 5, 6]$, $[7, 8, 9]$. The scalars 5.7, -1 (implicit), and $3/2 + i$ are called *coefficients* of the linear combination. By applying vector operations one may reduce the linear combination to the single vector

$$[87/10 + 7i, 72/5 + 8i, 201/10 + 9i].$$

11.2 Linear Independence

The function `reduce_lincomb` does this. It takes a list of coefficients and a list of vectors and returns the reduced linear combination.

```
-----------------------------------------------------------------
def reduce_lincomb(coeffs,vectors):
    lincomb = []
    for i in range(len(coeffs)):          # multiply coeff by vector
        c = coeffs[i]; v = vectors[i]
        lincomb.append(scalar_mult_vec(c,v))
    return add_vecs(lincomb)

-------------------------- Sample Run --------------------------
Input:
coeffs = '5.7, -1 ,3/3+i'.split(',')
vectors = tl.string2table('1,2,3; 4,5,6; 7,8,9')
print(reduce_lincomb(coeffs,vectors))
-----------------------------------------------------------------
Output:
['87/10+7i', '72/5+8i', '201/10+9i']
-----------------------------------------------------------------
```

11.2 Linear Independence

A set of n dimensional vectors is said to be *linearly dependent* if one or more of the vectors may be expressed as a linear combination of the others; otherwise, the vectors are said to be *linearly independent*. For example, the vectors

$$[1, 2, 3], \quad [4, 5, 6], \quad [7, 8, 9] \tag{11.1}$$

are linearly dependent, since

$$[7, 8, 9] = 2[4, 5, 6] - [1, 2, 3].$$

Note that one can write this equation as

$$x_1[1, 2, 3] + x_2[4, 5, 6] + x_3[7, 8, 9] = [0, 0, 0],$$

where $x_1 = x_3 = 1$ and $x_2 = -2$. We shall call an equation of this form a *dependency relation* among the vectors. If we apply vector operations we can further write this as

$$[x_1 + 4x_2 + 7x_3, \, 2x_1 + 5x_2 + 8x_3, \, 3x_1 + 6x_2 + 9x_3] = [0, 0, 0]$$

Equating components of these vectors we obtain the system

$$x_1 + 4x_2 + 7x_3 = 0$$
$$2x_1 + 5x_2 + 8x_3 = 0 \quad (11.2)$$
$$3x_1 + 6x_2 + 9x_3 = 0$$

Thus the linear dependence of the vectors in (11.1) is equivalent to the existence of a nontrivial, that is, not all zero, solution of the system (11.2).

The same analysis applies in general: m vectors

$$[a_{11}, a_{12}, \ldots, a_{1n}], [a_{21}, a_{22}, \ldots, a_{2n}], \ldots, [a_{m1}, a_{m2}, \ldots, a_{mn}] \quad (11.3)$$

are linearly dependent if the m by n system

$$a_{11}x_1 + a_{21}x_2 + \cdots + a_{mn}x_m = 0$$
$$a_{12}x_1 + a_{22}x_2 + \cdots + a_{m2}x_m = 0$$
$$\vdots \quad (11.4)$$
$$a_{1n}x_1 + a_{2n}x_2 + \cdots + a_{mn}x_m = 0$$

has a nontrivial solution. Notice that the columns of the coefficient matrix are the original vectors. The augmented matrix B of the system may be gotten by forming a matrix A with rows the vectors in (11.3), attaching a row of zeros, and taking the transpose of the result:

$$B = \begin{bmatrix} a_{11} & a_{12} & \cdots & a_{1n} \\ a_{21} & a_{22} & \cdots & a_{2n} \\ & & \vdots & \\ a_{m1} & a_{m2} & \cdots & a_{mn} \\ 0 & 0 & \cdots & 0 \end{bmatrix}^t = \begin{bmatrix} a_{11} & a_{21} & \cdots & a_{mn} & 0 \\ a_{12} & a_{22} & \cdots & a_{m2} & 0 \\ & & \vdots & & \\ a_{1n} & a_{2n} & \cdots & a_{mn} & 0 \end{bmatrix}$$

The *row rank* of a matrix is the largest number of linearly independent rows (considered as vectors). The function `get_lin_ind_rows` below returns a maximal set of linearly independent rows of a matrix and thus establishes its row rank. To see how the function works, consider the matrix

$$A = \begin{bmatrix} 1 & 2 & 3 & 4 & 5 \\ 6 & 7 & 8 & 9 & 10 \\ 11 & 12 & 13 & 14 & 15 \\ 16 & 17 & 18 & 19 & 20 \end{bmatrix}. \quad (11.5)$$

The first step is to determine if the first row of the matrix depends on the remaining rows. Thus we attempt to solve the system described by the equation

$$[1, 2, 3, 4, 5] = x_1[6, 7, 8, 9, 10] + x_2[11, 12, 13, 14, 15] + x_3[16, 17, 18, 19, 20],$$

11.2 Linear Independence

namely,

$$6x_1 + 11x_2 + 16x_3 = 1$$
$$7x_1 + 12x_2 + 17x_3 = 2$$
$$8x_1 + 13x_2 + 18x_3 = 3$$
$$9x_1 + 14x_2 + 19x_3 = 4$$
$$10x_1 + 15x_2 + 20x_3 = 5$$

The augmented matrix of the system is

$$\text{augmat} = \begin{bmatrix} 6 & 7 & 8 & 9 & 10 \\ 11 & 12 & 13 & 14 & 15 \\ 16 & 17 & 18 & 19 & 20 \\ 1 & 2 & 3 & 4 & 5 \end{bmatrix}^t = \begin{bmatrix} 6 & 11 & 16 & 1 \\ 7 & 12 & 17 & 2 \\ 8 & 13 & 18 & 3 \\ 9 & 14 & 19 & 4 \\ 10 & 15 & 20 & 5 \end{bmatrix}.$$

Using linsolve one determines that a solution x_1, x_2, x_3 exists, indicating that $[1, 2, 3, 4, 5]$ depends on the remaining rows. The next step is to determine whether the second row $[6, 7, 8, 9, 10]$ of A depends on $[11, 12, 13, 14, 15]$, and $[16, 17, 18, 19, 20]$. Thus we need to carry out the same procedure for the equation

$$[6, 7, 8, 9, 10] = x_1[11, 12, 13, 14, 15] + x_2[16, 17, 18, 19, 20].$$

The program determines that a solution x_1, x_2 exists, indicating that $[6, 7, 8, 9, 10]$ depends on the rows $[11, 12, 13, 14, 15]$ and $[16, 17, 18, 19, 20]$. The final step is to carry out the process for the equation

$$[11, 12, 13, 14, 15] = x_1[16, 17, 18, 19, 20].$$

This time the program determines that a solution does not exist. (One can also see this directly.) We conclude that the vectors $[11, 12, 13, 14, 15]$ and $[16, 17, 18, 19, 20]$ are linearly independent and the vectors $[1, 2, 3, 4, 5]$ and $[6, 7, 8, 9, 10]$ are linear combinations of these. The function then returns the former pair. Here is the code:

```
def get_lin_ind_rows(A):
    B = tl.copylist(A)                       # don't change matrix A
    while True:
        C = B[1:]+ [B[0]]                    # put first row on bottom
        augmat = ls.transpose(C)
        solution_list = ls.linsolve([],augmat,'c',False)
        if solution_list == []:              # no solution
            return B # return linearly independent vectors as matrix
        B = B[1:]    # otherwise remove top row and proceed again
```

It should be mentioned that the set of linear independent rows of a matrix is not unique. For example, one could carry out the above process starting instead from the last row and working up.

The function print_dependency_relation(A,LI) below takes a list of rows in the form of a matrix A and a sublist LI of linearly independent rows and returns a dependency relation for each row in A in LI. For our example, the function solves each of the systems derived from the following equations for x_1 and x_2

$$[1, 2, 3, 4, 5] = x_1[11, 12, 13, 14, 15] + x_2[16, 17, 18, 19, 20]$$
$$[6, 7, 8, 9, 10] = x_1[11, 12, 13, 14, 15] + x_2[16, 17, 18, 19, 20],$$

and returns these dependency relations, the vectors on the left thus expressed in terms of the linearly independent vectors on the right.

```
def print_dependency_relation(A,LI):
    for row in A:          # run through the rows of A that are not in LI
        if row in LI: continue
        B = LI+[row]                      # put row at bottom of LI
        augmat = ls.transpose(B)
        sol = ls.linsolve([],augmat,'x',False)
        print(row,'=')
        for i in range(len(sol)):
            coeff = sol[i].split('=')[1]
            if i < len(sol)-1:
                print('      ('+ coeff +')', LI[i],'+')
            else:
                print('      ('+ coeff +')', LI[i])
        print('\n')
```

-------------------------- Sample Run --------------------------
```
Input:
A = '1,2,3,4,5; 6,7,8,9,10; 11,12,13,14,15; 16,17,18,19,20'
A = tl.string2table(A)
LI = get_lin_ind_rows(A)
print(LI,'\n')
print_dependency_relation(A,LI)
```
```
Output:
['1', '2', '3', '4', '5'] = (3) ['11', '12', '13', '14', '15'] +
                            (-2) ['16', '17', '18', '19', '20']
['6', '7', '8', '9', '10'] = (2) ['11', '12', '13', '14', '15'] +
                             (-1) ['16', '17', '18', '19', '20']
```

11.3 The Range of a Matrix

Let V_n denote the collection of all n-dimensional vectors with Gaussian rational components. A nonempty subset V of V_n is said to be a *vector space* if sums and scalar multiples of members of V are again members of V, that is, if V is closed under the formation of linear combinations. Obviously, V_n is itself a vector space. A vector space V is said to be *spanned by vectors* $v_1, v_2, \ldots, v_p$ if it consists of all linear combinations of these vectors. In this case we call V the *span* of $v_1, v_2, \ldots, v_p$ and write $V = \text{vspan}\{v_1, v_2, \ldots, v_p\}$. For example, the span of $[1, 2, 3], [4, 5, 6], [7, 8, 6]$ consists of the linear combinations

$$x_1[1, 2, 3] + x_2[4, 5, 6] + x_3[7, 8, 6].$$

A linearly independent set of vectors that spans V is called a *basis for* V. It may be shown that every vector space has a basis and that any two bases have the same number of vectors. That number is called the *dimension* of the vector space.

There are two special vector spaces associated with an $m \times n$ matrix A, the range and the kernel. We treat the former in this section and the latter in the next. To describe these we use the notation Ax, where $x = [x_1, x_2, \ldots, x_n]$, for the transpose of the column vector obtained by multiplying A by the transpose of x. Thus if

$$A = \begin{bmatrix} 1 & 2 & 3 \\ 4 & 5 & 6 \\ 7 & 8 & 9 \end{bmatrix} \quad \text{and} \quad x = [x_1, x_2, x_3],$$

then Ax is the transpose of

$$\begin{bmatrix} 1 & 2 & 3 \\ 4 & 5 & 6 \\ 7 & 8 & 9 \end{bmatrix} \begin{bmatrix} x_1 \\ x_2 \\ x_3 \end{bmatrix} = \begin{bmatrix} x_1 + 2x_2 + 3x_3 \\ 4x_1 + 5x_2 + 6x_3 \\ 7x_1 + 8x_2 + 9x_3 \end{bmatrix}$$

that is,
$$Ax = [x_1 + 2x_2 + 3x_3, 4x_1 + 5x_2 + 6x_3, 7x_1 + 8x_2 + 9x_3]. \tag{11.6}$$

The *range* R of an $m \times n$ matrix A consists of all m dimensional vectors of the form Ax, where x is an arbitrary n dimensional vector. By properties of matrix multiplication, vector addition, and scalar multiplication, R is a vector space. Furthermore, since every vector x is a linear combination of the n dimensional vectors $[1, 0, 0, \ldots, 0], [0, 1, 0, \ldots, 0], \ldots, [0, 0, \ldots, 1]$, it follows that every vector Ax is a linear combination of the vectors

$$A[1, 0, 0, \ldots, 0], \ A[0, 1, 0, \ldots, 0], \ldots, A[0, 0, \ldots, 1].$$

These vectors are simply the columns of A. Thus we can find a basis for the range R by applying the function `get_lin_ind_rows` of the preceding section to the transpose of A:

```
def range_basis(A):
    At = ls.transpose(A)
    return get_lin_ind_rows(At)
```

--------------------------- Sample Run ---------------------------
```
Input:
A = '1,2,3,4,5; 6,7,8,9,10; 11,12,13,14,15; 16,17,18,19,20'
A = tl.string2table(A)
print(range_basis(A))
print(LI)
```

```
Output:
[['4', '9', '14', '19'], ['5', '10', '15', '20']]
```

11.4 The Kernel of a Matrix

The *kernel* of a matrix A is the set K of all n dimensional vectors x such that Ax is the zero vector. Properties of matrix multiplication, vector addition, and scalar multiplication show that K is a vector space. Finding a basis for K means extracting a maximal linearly independent set from the solutions x of the equation $Ax = 0$. We illustrate the method for the matrix

$$A = \begin{bmatrix} 1 & 2 & 3 & 4 & 5 \\ 6 & 7 & 8 & 9 & 10 \\ 11 & 12 & 13 & 14 & 15 \end{bmatrix},$$

for which K is the set of all solutions of the system

$$\begin{bmatrix} 1 & 2 & 3 & 4 & 5 \\ 6 & 7 & 8 & 9 & 10 \\ 11 & 12 & 13 & 14 & 15 \end{bmatrix} \begin{bmatrix} x_1 \\ x_2 \\ x_3 \\ x_4 \\ x_5 \end{bmatrix} = \begin{bmatrix} 0 \\ 0 \\ 0 \end{bmatrix}.$$

The system has augmented matrix

$$\begin{bmatrix} 1 & 2 & 3 & 4 & 5 & 0 \\ 6 & 7 & 8 & 9 & 10 & 0 \\ 11 & 12 & 13 & 14 & 15 & 0 \end{bmatrix}$$

The solution of the system given by `linsolve` is

```
['x1=3x5+2x4+x3', 'x2=-4x5-3x4-2x3', 'x3', 'x4', 'x5']
```

11.4 The Kernel of a Matrix

Here, x1,x2 are the dependent variables and x3,x4,x5 the free variables. Thus K consists of all vectors of the form

$$['x3+2x4+3x5', '-2x3-3x4-4x5', 'x3', 'x4', 'x5'].$$

We can get a basis for K by first setting x3=1,x4=0,x5=0, and then setting x3=0,x4=1,x5=0, and finally setting x3=0,x4=0,x5=1. This produces the vectors

$$[1,-2,1,0,0], [2,-3,0,1,0], [3,-4,0,0,1]|,$$

which form a basis. This is so because the linear independence of the "partial" vectors [1,0,0],[0,1,0],[0,0,1] confers linear independence on the full vectors

The following code implements the procedure. It takes the solution given by linsolve, detects the position of the first free variable, removes the dependent variables and their equality signs, and replaces the independent variables with 1's and 0's as described above.

```
def kernel_basis(A):
    zero_row = tl.zero_list(len(A[0]))
    At = ls.transpose(A)
    augmat = ls.transpose(At + [zero_row])
    sol_vec = ls.linsolve([],augmat,'x',False)
    free_vars = []
    for v in sol_vec:                      # get free variables
        for j in range(1,len(sol_vec)+1):
            var = 'x'+ str(j)
            if v == var:
                free_vars = free_vars + [v]
    basis = []
    sol_string = ','.join(sol_vec)   # condense to a string for ease
    for v in free_vars:              # run through free variables
        s = sol_string
        s = s.replace(v,'(1)')                       # substitute 1
        for w in free_vars:    # substitute 0 for other free vars
            if w != v:
                s = s.replace(w,'(0)')
        s = s.split(',')                      # make into a vector
        basis = basis+[s]
    for k in range(len(basis)):       # simplify members of basis
        for j in range(len(basis[k])):
            if '=' in basis[k][j]:
                right_side = basis[k][j].split('=')[1]
                basis[k][j] = ar.main(right_side)[0]
            else:
                basis[k][j] = ar.main(basis[k][j])[0]
    return basis
```

```
--------------------------- Sample Run ---------------------------
Input:
A = '1,2,3,4,5,6,7; 8,9,10,11,12,13,14; 15,16,17,18,19,20,21;\
     22,23,24,25,26,27,28'
A = tl.string2table(A3)
KB = kernel_basis(A)
tl.format_print(KB,2,'right')              # print basis vectors
print('\n')

#check that Au = zero list for all u in KB
for u in KB:
    print(mult_mat_vec(A,u))
------------------------------------------------------------------
Output:
1  -2  1  0  0  0  0                       # basis vectors
2  -3  0  1  0  0  0
3  -4  0  0  1  0  0
4  -5  0  0  0  1  0
5  -6  0  0  0  0  1

['0', '0', '0', '0']                       # Au = zero list
['0', '0', '0', '0']
['0', '0', '0', '0']
['0', '0', '0', '0']
['0', '0', '0', '0']
------------------------------------------------------------------
```

12 Determinants

The determinant of a square matrix A is a certain number calculated from the entries of A. In this chapter we describe and implement in Python several ways to find that number, and give some properties and applications. The module is headed by the import statements

```
------------------------Determinants.py ----------------------
import Tools as tl
import Arithmetic as ar
import LinSolve as ls
import MatAlg as mat
import Vectors as vec
import PolyAlg as pa
import MultiAlg as mu
---------------------------------------------------------------
```

12.1 Permutations

The standard definition of a determinant requires the notion of permutation and parity. A *permutation* of a set of symbols is an arrangement of the symbols. For simplicity, we shall always take the symbols to be integers $1, 2, \ldots, n$ for some n, called the *length* of the partition. We denote a permutation by a list and the collection of permutations by a list of these lists. For example, the list of length three permutations is

```
[[1, 2, 3], [1, 3, 2], [2, 1, 3], [2, 3, 1], [3, 1, 2], [3, 2, 1]]
```

The number of permutations of n symbols is $n!$, since there are n different ways to choose the first symbol of the permutation, and for each of those $n-1$ ways to choose the second symbol, resulting in $n(n-1)$ ways to choose the first two symbols, etc.

The function permutations(n) returns the list of permutations of length n. It does so by first creating a list of single number lists, then attaching a distinct second number to each of the single number list to form a list of double number lists, etc. The process continues until the desired list of permutations is obtained. In the code we have included print statements that illuminate the process.

```
def permutations(n):                           # generates length n perms
    perms = [[k] for k in range(1,n+1)]        # first list (singles)
    perm_lists = [perms]                       # put in list of perms
    print(perms)                               # first step
    for i in range(1,n):                       # generate higher order lists
        previous_perms = perm_lists[i-1]
        new_perms = []
        for j in range(1,n+1):                 # attach symbols to previous perms
            for k in range(len(previous_perms)):
                if j not in previous_perms[k]:
                    new_perms.append([j]+previous_perms[k])
        perm_lists.append(new_perms)           # next higher length perm
        print(new_perms)                       # print newest perm list
    return perm_lists[len(perm_lists)-1]       # last: the desired list
```

```
-------------------------- Sample Run --------------------------
Input:
permutations(3)
----------------------------------------------------------------
Output:
[[1], [2], [3]]
[[1, 2], [1, 3], [2, 1], [2, 3], [3, 1], [3, 2]]
[[1, 2, 3], [1, 3, 2], [2, 1, 3], [2, 3, 1], [3, 1, 2], [3, 2, 1]]
----------------------------------------------------------------
```

A permutation p is said to be *odd*, respectively, *even*, if the number of pairs i, j in the permutation with $i < j$ and $p[i] > p[j]$ odd, respectively, even. The *sign* of p is -1 is the permutation is odd, and $+1$ if the permutation is even. The following function returns the sign of a permutation by counting the number of "out of order" pairs.

```
def permutation_sign(p):
    L = len(p)
    parity = 0
    for i in range(L):
        for j in range(i+1,L):
            if p[j] < p[i]:
                parity += 1                    # i < j, p[j] < p[i]
    return (-1)**parity
```

```
------------------------ Sample Run ------------------------
Input:
p = [3,0,2,4,1]
q = [3,0,2,1,4]
print(permutation_sign(p),',', permutation_sign(q))
------------------------------------------------------------
Output:
-1,1
------------------------------------------------------------
```

12.2 Leibniz Formula for a Determinant

Let A be a matrix of size n, say

$$A = \begin{bmatrix} a_{11} & a_{12} & \cdots & a_{1n} \\ a_{21} & a_{22} & \cdots & a_{2n} \\ \vdots & \vdots & \ddots & \vdots \\ a_{n1} & a_{n2} & \cdots & a_{nn} \end{bmatrix}.$$

The *determinant of* A, denoted by $|A|$, or $\det(A)$ is defined as the sum of all terms $s(p)a_{1p[1]}a_{2p[2]}\ldots,a_{np[n]}$, where p ranges through the permutations of p of the subscripts $1, 2, \ldots, n$ and $s(p)$ is the sign of p. We write this explicitly as

$$\begin{vmatrix} a_{11} & a_{12} & \cdots & a_{1n} \\ a_{21} & a_{22} & \cdots & a_{2n} \\ \vdots & \vdots & \ddots & \vdots \\ a_{n1} & a_{n2} & \cdots & a_{nn} \end{vmatrix} = \sum_{p \in P} s(p)a_{1p[1]}a_{1p[2]}\ldots,a_{np[n]},$$

where P denotes the set of all permutations of $1, 2, \ldots, n$. Here is a function that calculates the determinant of a square matrix A using the Leibniz formula. As usual, we take the entries taken be Gaussian rational numbers.

```
------------------------------------------------------------
def det_leibnitz(A):
    n = len(A[0])    # determinant size
    perms = permutations(n)
    d = '0'
    for p in perms:
        d = ar.main(d + '+(' + perm_term(A,p) + ')')[0]
    return d
------------------------------------------------------------
```

12.3 Laplace Expansion of a Determinant

The Laplace method of evaluating the determinant is recursive in that it depends on determinants of matrices of lesser size, ultimately reaching the case $n = 2$:

$$\begin{vmatrix} a & b \\ c & d \end{vmatrix} = ad - bc. \tag{12.1}$$

For $n > 2$ the value may be found by expansion along any row or column. We illustrate by expanding along row 1 for the case $n = 3$:

$$\begin{vmatrix} a_{11} & a_{12} & a_{13} \\ a_{21} & a_{22} & a_{23} \\ a_{31} & a_{32} & a_{33} \end{vmatrix} = a_{11} \begin{vmatrix} a_{22} & a_{23} \\ a_{32} & a_{33} \end{vmatrix} - a_{12} \begin{vmatrix} a_{21} & a_{23} \\ a_{31} & a_{33} \end{vmatrix} + a_{13} \begin{vmatrix} a_{21} & a_{22} \\ a_{31} & a_{32} \end{vmatrix}.$$

The terms in the expansion are the entries of row 1, with alternating signs, multiplied by second order determinants. These are obtained from the original matrix by deleting row 1 and, successively, the columns of the row 1 entries. Here's an example:

$$\begin{vmatrix} 1 & 2 & 3 \\ 4 & 5 & 6 \\ 7 & 8 & 9 \end{vmatrix} = 1 \cdot \begin{vmatrix} 5 & 6 \\ 8 & 9 \end{vmatrix} - 2 \cdot \begin{vmatrix} 4 & 6 \\ 7 & 9 \end{vmatrix} + 3 \cdot \begin{vmatrix} 4 & 5 \\ 7 & 8 \end{vmatrix} = -3 - 2(-6) + 3(-3) = 0.$$

Any row may be used for expansion. Column expansions may also be used. The general rule consists of multiplying an entry a_{ij} by $(-1)^{i+j}$ times the determinant obtained by deleting row i and column j. In evaluating a determinant one typically picks the row or column with the most zeros so as to minimize the number of multiplications. We illustrate with the following example, evaluating along first columns.

$$\begin{vmatrix} 1 & 2 & 3 & 4 \\ 0 & 5 & 6 & 7 \\ 0 & 0 & 8 & 9 \\ 0 & 0 & 0 & 10 \end{vmatrix} = 1 \cdot \begin{vmatrix} 5 & 6 & 7 \\ 0 & 8 & 9 \\ 0 & 0 & 10 \end{vmatrix} = 1 \cdot 5 \cdot \begin{vmatrix} 8 & 9 \\ 0 & 10 \end{vmatrix} = 1 \cdot 5 \cdot 8 \cdot 10 = 400.$$

The matrices in the calculations are said to be *upper triangular*, having zeros below the main diagonal. The above calculations show that the determinant of such a matrix is simply the product of the entries along the main diagonal.

Here's a recursive function that uses the first row of the matrix to evaluate a determinant.

```
def det_laplace(A):
    n = len(A)
    if n == 2:                                              # base case
        return ar.main( '('+ A[0][0] + ')('+ A[1][1] +')-  \
                        ('+ A[1][0] + ')('+ A[0][1] +')')[0]
```

12.3 Laplace Expansion of a Determinant

```
        d = '0'
        for k in range(n):
            B = remove_row_col(A,0,k)
            sign = str((-1)**k)
            d = ar.main(d + '+('+ sign +')('+ A[0][k] +') \
                                ('+ det_laplace(B) + ')')[0]
        return d

    def remove_row_col(A,r,c):
        # returns A with row r and column c deleted
        n = len(A) # size of determinant
        B = []; C = []; D = []
        for k in range(n):                    # append all rows except r
            if k != r:
                B.append(A[k])
        C = ls.transpose(B)
        for k in range(n):                    # append all rows except c
            if k != c:
                D.append(C[k])
        return ls.transpose(D)
```

The following version of det_laplace allow letters as determinant entries. These may serve in applications as parameters or variables. It is gotten by replacing ar.main in the above code by mu.main. The function will be useful later in calculating the characteristic polynomial of a matrix.

```
    def det_params(A):
        n = len(A)
        if n == 2:
            return mu.main( '('+ A[0][0] + ')('+ A[1][1] +')- \
                            (('+ A[1][0] +')('+ A[0][1] +'))')[0]
        d = '0'
        for k in range(n):
            B = remove_row_col(A,0,k)
            sign = str((-1)**k)
            d = mu.main(d + '+('+ sign +')('+ A[0][k] +') \
                ('+ det_params(B) + ')')[0]
        return d

    ---------------------------- Sample Run ---------------------------
    Input:
    A = tl.string2table('1-x,1,2y;3,2-ax,-1;-1,2,3-xz')
    print(det_params(A))
    ---------------------------------------------------------------
    Output:
    -ax^3z+ax^2z-2axy+2x^2z+3ax^2-3ax+xz+16y-8x
    ---------------------------------------------------------------
```

12.4 Properties of Determinants

The following properties are useful several contexts, including evaluating determinants. The reader will notice the connection of the first three properties with row operations on a matrix. We exploit this connection in the next section.

- Switching a pair of rows (or columns) changes the sign of the determinant.
- Multiplying a row or column by a scalar k multiplies the determinant by k. Thus rows or columns may be factored.
- Adding a multiple of one row to another does not change the value of the determinant.
- The determinant of a matrix is the same as the determinant of its transpose.
- The determinant of the product of matrices is the product of the determinants.
- The determinant of the inverse of a matrix is the reciprocal of the determinant of the matrix.

Here's an example that uses the first three properties to evaluate a determinant. The method is akin to the row echelon algorithm.

$$\begin{vmatrix} 2 & 9 & 3 \\ 2 & 6 & 0 \\ 3 & 14 & -5 \end{vmatrix} \xrightarrow{r1<->r2} (-1)\begin{vmatrix} 2 & 6 & 0 \\ 2 & 9 & 3 \\ 3 & 14 & -5 \end{vmatrix} = (-2)\begin{vmatrix} 1 & 3 & 0 \\ 2 & 9 & 3 \\ 3 & 14 & -5 \end{vmatrix}$$

$$\xrightarrow{\substack{-2r1+r2 \\ -3r1+r3}} (-2)\begin{vmatrix} 1 & 3 & 0 \\ 0 & 3 & 3 \\ 0 & 5 & -5 \end{vmatrix} = (-2)\cdot 3 \cdot 5 \begin{vmatrix} 1 & 3 & 0 \\ 0 & 1 & 1 \\ 0 & 1 & -1 \end{vmatrix}$$

$$\xrightarrow{-1r2+r3} (-30)\begin{vmatrix} 1 & 3 & 0 \\ 0 & 1 & 1 \\ 0 & 0 & -2 \end{vmatrix} = 60$$

The last calculation uses the fact, as noted earlier, that the determinant of an upper triangular matrix is the product of the entries along the main diagonal.

12.5 Determinants Using Row Echelon

The function `det_echelon(A)` uses the properties in the preceding section to evaluate the determinant of A. It does so by running `ls.row_echelon(A)`, which, recall, produces not only the reduced row echelon form R of A but also the number `switches` of row switches and the product `prod` of the matrix entries whose reciprocals are the scalar multipliers in type 2 operations during the pivoting process. Since R is an upper triangular matrix, det(R) is the product of its diagonal elements. Multiplying det(R) by `prod` and by `(-1)^switches` undoes the effects of row operations on A and so produces det(A).

```
def det_echelon(A):
    n = len(A[0])                           #size of determinant
    R = ls.row_echelon(A)[0]
    d = R[0][0]                             # first diagonal entry
    for k in range(1,n):                    # get product of diagonal entries
        d =   ar.main( '('+ d +')('+ R[k][k] +')')[0]
    s = str((-1)**ls.switches)
    p = ls.prod
    return ar.main('('+ d +')('+ s +')('+ p +')')[0]
```

The echelon method of evaluating a determinant is far superior to the previous methods considered. For example, using `det_echelon(A)` on the matrix

$$A = \begin{vmatrix} 0 & 0 & 0 & 0 & 0 & 0 & 0 & 0 & 0 & 0 & 1 \\ 0 & 0 & 0 & 0 & 0 & 0 & 0 & 0 & 0 & 2 & 0 \\ 0 & 0 & 0 & 0 & 0 & 0 & 0 & 0 & 3 & 0 & 0 \\ 0 & 0 & 0 & 0 & 0 & 0 & 0 & 4 & 0 & 0 & 0 \\ 0 & 0 & 0 & 0 & 0 & 0 & 5 & 0 & 0 & 0 & 0 \\ 0 & 0 & 0 & 0 & 0 & 6 & 0 & 0 & 0 & 0 & 0 \\ 0 & 0 & 0 & 0 & 7 & 0 & 0 & 0 & 0 & 0 & 0 \\ 0 & 0 & 0 & 8 & 0 & 0 & 0 & 0 & 0 & 0 & 0 \\ 0 & 0 & 9 & 0 & 0 & 0 & 0 & 0 & 0 & 0 & 0 \\ 0 & 10 & 0 & 0 & 0 & 0 & 0 & 0 & 0 & 0 & 0 \\ 11 & 0 & 0 & 0 & 0 & 0 & 0 & 0 & 0 & 0 & 0 \end{vmatrix}$$

produces the value -39916800 almost instantaneously while the other methods basically give up (at least on the author's machine).

12.6 Cramer's Rule

Determinants provide another way to solve $n \times n$ systems of equations

$$\begin{aligned} a_{11}x_1 + a_{12}x_2 + \cdots + a_{1n}x_n &= b_1 \\ a_{21}x_1 + a_{22}x_2 + \cdots + a_{2n}x_n &= b_2 \\ &\vdots \\ a_{n1}x_1 + a_{n2}x_2 + \cdots + a_{nn}x_n &= b_n. \end{aligned} \quad (12.2)$$

To describe the method, write the system in matrix form as $AX = B$ and let A_k be the matrix obtained by replacing column k by the column B:

$$A_k = \begin{bmatrix} a_{11} & \cdots & a_{1,k-1} & b_1 & a_{1,k+1} & \cdots & a_{1n} \\ a_{21} & \cdots & a_{2,k-1} & b_2 & a_{2,k+1} & \cdots & a_{2n} \\ \vdots & & \vdots & \vdots & \vdots & & \vdots \\ a_{n1} & \cdots & a_{n,k-1} & b_n & a_{n,k+1} & \cdots & a_{nn} \end{bmatrix}.$$

Cramer's rule asserts that if $\det(A) \neq 0$, then the system (12.2) has the unique solution $(x_1, \ldots, x_n)$, where $x_k = |A_k|/|A|$. The method is readily implemented in Python:

```
def cramer_rule(A,B):
    S = []                                      # for solution list
    d = det_echelon(A)
    if d == '0': return []
    for k in range(len(A[0])):
        C = replace_col(A,B,k)                  # replace column k of A with B
        c = det_echelon(C)
        ratio = ar.main('('+c+')/'+'('+d+')')[0]        #  c/d
        print('x'+str(k+1)+'=',ratio)           # print equations
        S.append(ratio)
    return S

def replace_col(A,B,col):
    n = len(A) # size of determinant
    C = []
    At = ls.transpose(A)
    for k in range(n):
        if k < col:                             # keep column k of A
            C.append(At[])
        if k == col:                            # replace column col with B
            C.append(B)
        if k > col and k < n:                   # keep column k of A
            C.append(At[k])
    return ls.transpose(C)
```

```
-------------------------- Sample Run --------------------------
Input:
A = '1,2,3;5,-3.1-i,1;-1,5,6+2i'; B = '4,7,8'
A = tl.string2table(A)
B = tl.string2list(B)
print('\n',cramer_rule(A,B))
----------------------------------------------------------------
Output:
x1= 141106/155945+(182512/155945)i
x2= 46/31189+(45212/31189)i
x3= 160738/155945-(211544/155945)i

['141106/155945+(182512/155945)i', '46/31189+(45212/31189)i',
 '160738/155945-(211544/155945)i']
----------------------------------------------------------------
```

12.6 Cramer's Rule

The function `cramer_rule_params(A,B)` is a version of `cramer_rule(A,B)` that allows parameters or variables in the matrices A or B. The main difference is that it uses `det_params` instead of `det_echelon` to calculate determinants. Equations with parameters are particularly useful in modelling situations that occur in areas such as physics and economics where one might need to see how solutions are affected by changing some of the specifics of the model, these represented by the parameters. The sample run for the case of a single parameter shows how the parameter shows up in the solution and how a slight variation of the parameter affects solution values.

```
----------------------------------------------------------------
def cramer_rule_params(A,B):
    S = []                                         # for solution
    d = det_params(A)
    if d == '0': return []
    for k in range(len(A[0])):
        C = replace_col(A,B,k)
        c = det_params(C)
        ratio = '('+ c +')/'+'('+ d +')'
        print('x'+str(k+1)+'=',ratio)
        S.append(ratio)
return S                                           # solution list

def evaluate_cramer(S,substitutions,p):            # p = decimal places
    for k in range(len(S)):
        e = mu.evaluate(S[k],substitutions,p)
        print('x'+str(k+1)+' =',e)

-------------------------- Sample Run --------------------------
Input:
A = '-3,2a,-5;-1,0,-2;3.09876,-4,1'
B = '1,2,3+a'
A = tl.string2table(A)
B = tl.string2list(B)
S = cramer_rule_parameter(A,B)
print('\n')
print('a = 1')
evaluate_cramer(S,'1',5)
print('\n')
print('a = 1.0001:')
evaluate_cramer(S,'1.0001',5)
----------------------------------------------------------------
Output:
x1= (-4a^2-16a+32)/((-64969/6250)a+4)
x2= (-a+52469/3125)/((-64969/6250)a+4)
x3= (2a^2+(114969/6250)a-20)/((-64969/6250)a+4)

a = 1:
x1 = -1.87646
x2 = -2.46912
x3 = .-6178
```

```
a = 1.0001:
x1 = -1.87578
x2 = -2.4687
x3 = .-6212
```

12.7 Application: Common Root of Polynomials

Determinants may be used to discover whether a two polynomials

$$a_m x^m + a_{m-1} x^{m-1} + \cdots + a_1 x + a_0, \quad b_n x^n + b_{n-1} x^{n-1} + \cdots + b_1 x + b_0, \quad (12.3)$$

have a common root. Consider the special case $m = n = 2$. As a first step we observe that if each equation has a root r, then, by long division of polynomials, we obtain

$$a_2 x^2 + a_1 x + a_0 = (x - r)(c_1 x + c_0) \text{ and } b_2 x^2 + b_1 x + b_0 = (x - r)(d_1 x + d_0)$$

for some constants c_i and d_i. Solving each equation for $x - r$ and equating the result we have

$$\frac{a_2 x^2 + a_1 x + a_0}{c_1 x + c_0} = x - r = \frac{b_2 x^2 + b_1 x + b_0}{d_1 x + d_0}$$

and so

$$(a_2 x^2 + a_1 x + a_0)(d_1 x + d_0) = (b_2 x^2 + b_1 x + b_0)(c_1 x + c_0).$$

Multiplying and collecting terms on each side of the last equation yields

$$a_2 d_1 x^3 + (a_1 d_1 + a_2 d_0) x^2 + (a_0 d_1 + a_1 d_0) x + a_0 d_0$$
$$= b_2 c_1 x^3 + (b_1 c_1 + b_2 c_0) x^2 + (b_0 c_1 + b_1 c_0) x + b_0 c_0.$$

Since this holds for all x, the coefficients of like powers of x must be equal:

$$a_2 d_1 = b_2 c_1, \; a_1 d_1 + a_2 d_0 = b_1 c_1 + b_2 c_0, \; a_0 d_1 + a_1 d_0 = b_0 c_1 + b_1 c_0, \; a_0 d_0 = b_0 c_0.$$

We regard this as a system of four linear equations in the four unknowns $d_1, d_0, -c_1$ and $-c_0$:

$$\begin{aligned} a_2 d_1 \phantom{{}+a_1 d_0} &+ \phantom{b_0(-c_1)+{}} b_2(-c_1) \phantom{{}+b_1(-c_0)} = 0 \\ a_1 d_1 + a_2 d_0 &+ b_1(-c_1) + b_2(-c_0) = 0 \\ a_0 d_1 + a_1 d_0 &+ b_0(-c_1) + b_1(-c_0) = 0 \\ a_0 d_0 \phantom{{}+a_1 d_0} &\phantom{{}+b_0(-c_1)} + b_0(-c_0) = 0 \end{aligned}$$

By Cramer's rule this has a nontrivial solution if and only if the determinant of the system is zero:

12.7 Application: Common Root of Polynomials

$$\begin{vmatrix} a_2 & 0 & b_2 & 0 \\ a_1 & a_2 & b_1 & b_2 \\ a_0 & a_1 & b_0 & b_1 \\ 0 & a_0 & 0 & b_0 \end{vmatrix} = 0$$

Taking transposes we obtain the equivalent condition

$$\begin{vmatrix} a_2 & a_1 & a_0 & 0 \\ 0 & a_2 & a_1 & a_0 \\ b_2 & b_1 & b_0 & 0 \\ 0 & b_2 & b_1 & b_0 \end{vmatrix} = 0,$$

the rows comporting nicely with the standard notation for polynomials.

One shows in a similar manner that polynomials (12.3), for the case $m = n$, have a common root if and only if

$$\begin{vmatrix} a_m & a_{m-1} & \cdots & a_1 & a_0 & 0 & \cdots & 0 \\ 0 & a_m & a_{m-1} & \cdots & a_1 & a_0 & \cdots & 0 \\ \vdots & \vdots & \vdots & \vdots & \vdots & \vdots & & 0 \\ 0 & \cdots & 0 & a_m & a_{m-1} & \cdots & a_1 & a_0 \\ b_m & b_{m-1} & \cdots & b_1 & b_0 & 0 & \cdots & 0 \\ 0 & b_m & b_{m-1} & \cdots & b_1 & b_0 & \cdots & 0 \\ \vdots & \vdots & \vdots & \vdots & \vdots & \vdots & & 0 \\ 0 & \cdots & 0 & b_m & b_{m-1} & \cdots & b_1 & b_0 \end{vmatrix} = 0.$$

For the case $m \neq n$ we can prepend zero terms to the polynomial with lowest degree. For example, for $m = 2$ and $n = 3$ we can write the quadratic (contrary to convention) as $0x^3 + a_2x^2 + a_1x + a_0 = 0$. The resulting determinant is

$$\begin{vmatrix} 0 & a_2 & a_1 & a_0 & 0 \\ 0 & 0 & a_2 & a_1 & a_0 \\ b_3 & b_2 & b_1 & b_0 & 0 \\ 0 & b_3 & b_2 & b_1 & b_0 \end{vmatrix} = 0,$$

Here is a program that takes as input two polynomials P, Q and returns True if they have a common zero and False otherwise.

```
def has_common_root(P,Q):
    listP = pa.main(P)[1]          # get the flists of the polynomials
    listQ = pa.main(Q)[1]
    M = make_mat(listP,listQ)      # make the matrix
    tl.format_print(M, 2, 'right') # display matrix
    return det_echelon(M) == '0'

def make_mat(listP,listQ):
```

```
        M = []
        LP = len(listP), LQ = len(listQ)
        if LP < LQ:                             # prepend zeros to listP
            listP = tl.zero_list(LQ - LP) + listP
        if LQ < LP:                             # prepend zeros to listQ
            listQ = tl.zero_list(LP - LQ) + listQ
        LP = len(listP); LQ = len(listQ)        # adjust values
        for j in range(LP):                     # make top half of matrix
            row = tl.zero_list(j) + listP + tl.zero_list(LQ-1-j)
            M.append(row)
        for j in range(LQ):                     # make bottom half of matrix
            row = tl.zero_list(j) + listQ + tl.zero_list(LP-1-j)
            M.append(row)
        return M

    --------------------------- Sample Run ---------------------------
    P = 'x^2 + 2x + 3';   Q = '4x^3 + 5x^2 + 6x + 7'
    print(has_common_root(P,Q),'\n')
    P = 'x^2-2x+1';   Q = 'x^3-3x^2+3x-1'
    print(has_common_root(P,Q))
    ------------------------------------------------------------------
    Output:
    0 1 2 3 0 0 0         0  1 -2  1  0  0  0
    0 0 1 2 3 0 0         0  0  1 -2  1  0  0
    0 0 0 1 2 3 0         0  0  0  1 -2  1  0
    0 0 0 0 1 2 3         0  0  0  0  1 -2  1
    4 5 6 7 0 0 0         1 -3  3 -1  0  0  0
    0 4 5 6 7 0 0         0  1 -3  3 -1  0  0
    0 0 4 5 6 7 0         0  0  1 -3  3 -1  0
    0 0 0 4 5 6 7         0  0  0  1 -3  3 -1
    False                 True
    ------------------------------------------------------------------
```

12.8 Application: Plane Through Three Points

A plane in a three dimensional xyz coordinate system has equation of the form $ax + by + cz + d = 0$, where the constants a, b, and c are not all zero. Four points (x, y, z), (x_1, y_1, z_1), (x_2, y_2, z_2), (x_3, y_3, z_3) in space are said to be *co-planar* if they lie on the same plane, that is, if there exist a, b, c, d, with a, b, and c not all zero, such that

$$ax + by + cz + d = 0$$
$$ax_1 + by_1 + cz_1 + d = 0$$
$$ax_2 + by_2 + cz_2 + d = 0$$
$$ax_3 + by_3 + cz_3 + d = 0$$

12.8 Application: Plane Through Three Points

This may be viewed as a system of linear equations in the variables a, b, c, d. It has a nontrivial solution if and only if

$$\begin{vmatrix} x & y & z & 1 \\ x_1 & y_1 & z_1 & 1 \\ x_2 & y_2 & z_2 & 1 \\ x_3 & y_3 & z_3 & 1 \end{vmatrix} = 0.$$

Thus if three given points (x_1, y_1, z_1), (x_2, y_2, z_2), and (x_3, y_3, z_3) do not lie on the same line, then a variable point (x, y, z) lies on the plane if and only if the previous determinant equality holds. Expanding the determinant along the first row yields the alternate form $Ax - By + Cz = D$, where

$$A = \begin{vmatrix} y_1 & z_1 & 1 \\ y_2 & z_2 & 1 \\ y_3 & z_3 & 1 \end{vmatrix}, \quad B = \begin{vmatrix} x_1 & z_1 & 1 \\ x_2 & z_2 & 1 \\ x_3 & z_3 & 1 \end{vmatrix}, \quad C = \begin{vmatrix} x_1 & y_1 & 1 \\ x_2 & y_2 & 1 \\ x_3 & y_3 & 1 \end{vmatrix}, \quad D = \begin{vmatrix} x_1 & y_1 & z_1 \\ x_2 & y_2 & z_2 \\ x_3 & y_3 & z_3 \end{vmatrix}.$$

Here is a function that takes three points given as the Python lists P1 = [x1,y1,z1], P2 = [x2,y2,z2], and P3 = [x3,y3,z3], and returns the equation of the plane through these points.

```
def plane(P1,P2,P3):
    A = [[P1[1],P1[2],'1'],  [P2[1],P2[2],'1'],  [P3[1],P3[2],'1'] ]
    B = [[P1[0],P1[2],'1'],  [P2[0],P2[2],'1'],  [P3[0],P3[2],'1'] ]
    C = [[P1[0],P1[1],'1'],  [P2[0],P2[1],'1'],  [P3[0],P3[1],'1'] ]
    D = [[P1[0],P1[1],P1[2]],[P2[0],P2[1],P2[2]],[P3[0],P3[1],P3[2]]]
    A = det_echelon(A)
    B = det_echelon(B)
    C = det_echelon(C)
    D = det_echelon(D)
    A = tl.add_parens(A)
    B = tl.add_parens(B)
    C = tl.add_parens(C)
    D = tl.add_parens(D)
    eqn = A + 'x' ' - ' + B + 'y' ' + ' + C + 'z' ' = ' + D
    return tl.fix_signs(eqn)
```

--------------------------- Sample Run ---------------------------
```
Input:
P1 ='1,0,-3'.split(',')
P2 ='0,-2,3'.split(',')
P3 ='1,-4,0'.split(',')
print(plane(P1,P2,P3
```

```
Output:
18x+3y+4z=6
```

12.9 Application: Sphere Through Four Points

A sphere with center (a, b, c) and radius r is given by the equation

$$(x - a)^2 + (y - b)^2 + (z - c)^2 = r^2, \qquad (12.4)$$

where (x, y, z) are the coordinates of a general point in space. Given four points (x_1, y_1, z_1), (x_2, y_2, z_2), (x_3, y_3, z_3), and (x_4, y_4, z_4) that do not lie in the same plane, there exists a unique sphere containing these points. This is seen by solving the following system of equations for a, b, c and r:

$$(x_1 - a)^2 + (y_1 - b)^2 + (z_1 - c)^2 = r^2$$
$$(x_2 - a)^2 + (y_2 - b)^2 + (z_2 - c)^2 = r^2$$
$$(x_3 - a)^2 + (y_3 - b)^2 + (z_3 - c)^2 = r^2$$
$$(x_4 - a)^2 + (y_4 - b)^2 + (z_4 - c)^2 = r^2$$

These are *nonlinear* equations, but the system may be linearized by expanding the equations and rewriting the system. For example, expanding the first equation gives

$$x_1^2 - 2x_1 a + a^2 + y_1^2 - 2y_1 b + b^2 + z_1^2 - 2z_1 c + c^2 = r^2,$$

which may be written

$$2x_1 a + 2y_1 b + 2z_1 c + r^2 - a^2 - b^2 - c^2 = x_1^2 + y_1^2 + z_1^2.$$

Thus, setting $k = r^2 - a^2 - b^2 - c^2$, we may write the above system as

$$2x_1 a + 2y_1 b + 2z_1 c + k = x_1^2 + y_1^2 + z_1^2$$
$$2x_2 a + 2y_2 b + 2z_2 c + k = x_2^2 + y_2^2 + z_2^2$$
$$2x_3 a + 2y_3 b + 2z_3 c + k = x_3^2 + y_3^2 + z_3^2$$

This is a linear system which may be solved for the unknowns a, b, c, k, and consequently for r, by using Cramer's rule on the matrices

$$A = \begin{bmatrix} 2x_1 & 2y_1 & 2z_1 & 1 \\ 2x_2 & 2y_2 & 2z_2 & 1 \\ 2x_3 & 2y_3 & 2z_3 & 1 \end{bmatrix} \quad \text{and} \quad B = \begin{bmatrix} x_1^2 + y_1^2 + z_1^2 \\ x_2^2 + y_2^2 + z_2^2 \\ x_3^2 + y_3^2 + z_3^2 \end{bmatrix}.$$

The function `sphere(P1,P2,P3,P4)` implements this process. It takes as input the four points, written as lists of coordinates x_i, y_i, z_i, and outputs the center and the radius of the unique sphere through these points.

12.9 Application: Sphere Through Four Points

```
def sphere(P1,P2,P3,P4):
    A = []; B = []
    x1 = P1[0]; y1 = P1[1]; z1 = P1[2]
    x2 = P2[0]; y2 = P2[1]; z2 = P2[2]
    x3 = P3[0]; y3 = P3[1]; z3 = P3[2]
    x4 = P4[0]; y4 = P4[1]; z4 = P4[2]

    a11 = ar.main('2('+ x1 +')')[0]
    a12 = ar.main('2('+ y1 +')')[0]
    a13 = ar.main('2('+ z1 +')')[0]
    A.append([a11,a12,a13,'1'])

    a21 = ar.main('2('+ x2 +')')[0]
    a22 = ar.main('2('+ y2 +')')[0]
    a23 = ar.main('2('+ z2 +')')[0]
    A.append([a21,a22,a23,'1'])

    a31 = ar.main('2('+ x3 +')')[0]
    a32 = ar.main('2('+ y3 +')')[0]
    a33 = ar.main('2('+ z3 +')')[0]
    A.append([a31,a32,a33,'1'])

    a41 = ar.main('2('+ x4 +')')[0]
    a42 = ar.main('2('+ y4 +')')[0]
    a43 = ar.main('2('+ z4 +')')[0]
    A.append([a41,a42,a43,'1'])


    b1 = ar.main(x1+'^2+' + y1 +'^2+' + z1 +'^2')[0]
    b2 = ar.main(x2+'^2+' + y2 +'^2+' + z2 +'^2')[0]
    b3 = ar.main(x3+'^2+' + y3 +'^2+' + z3 +'^2')[0]
    b4 = ar.main(x4+'^2+' + y4 +'^2+' + z4 +'^2')[0]

    B = [b1,b2,b3,b4]
    C = cramer_rule(A,B)
    if C == []:
        print('no sphere')
        return '',''

    a,b,c,k = C[0], C[1], C[2], C[4]
    d = '('+ a +')^2+' + '('+ b +')^2+' + '('+ c +')^2'
    rsquared = ar.main(k + '+' + d)[0]
    radius = '('+ rsquared + ')^(1/2)'
    center = [a,b,c]
    return center,radius
```

-------------------------- Sample Run --------------------------
Input:
P1 ='1,0,3'.split(',')
P2 ='0,2,3'.split(',')
P3 ='1,4,0'.split(',')
P4 ='1,2,5'.split(',')

```
center, radius = sphere(P1,P2,P3,P4)
print('center   = ',center)
print('radius   = ',radius)
```

```
Output:
center   =  ['53/14', '37/14', '33/14']
radius   =  (2971/196)^(1/2)
```

12.10 Eigenvalues and Eigenvectors

An *eigenvalue* of an $n \times n$ matrix A is a number x such that $AX = xX$ or, equivalently, $(A - xI)X = 0$, for some nonzero vector X. The vector X is then called an *eigenvector of* A associated with the eigenvalue x. The *eigenspace* of A associated with the eigenvalue x is the zero vector together all eigenvectors. Thus the eigenspace is the kernel of the matrix $(A - xI)$.

By Cramer's rule the equation $(A - xI)X = 0$ has a non trivial solution X if and only if $\det(A - xI) = 0$. The latter equation is called the *characteristic equation* of A and the determinant the *characteristic polynomial* of A. Eigenvalues have important applications in several disciplines, particularly in physics and engineering.

The following program returns the characteristic polynomial of A. It uses the function det_params of Sect. 12.3.

```
def char_pol(A):
    B = tl.string2table(A)
    for i in range(len(B)):
        B[i][i] = '(' + B[i][i] + '-x)'   # subtract x from diagonal
    return det_params(B)
```

```
------------------------- Sample Run ---------------------------
Input:
A = '1+i,2-3i,3;4,5,6;7,8-5i,9'
print(char_pol(A))
```

```
Output:
-x^3+(15+i)x^2+(18-56i)x+(-30-51i)
```

The *Cayley-Hamilton theorem* asserts that substituting A into its characteristic polynomial results in the zero matrix. For example, the characteristic polynomial of $A = \begin{bmatrix} 1 & 2 \\ 3 & 4 \end{bmatrix}$ is

12.11 Adjugate Matrix

$$\begin{vmatrix} 1-x & 2 \\ 3 & 4-x \end{vmatrix} = x^2 - 5x - 2$$

Substituting A into this (replacing 2 by 2I) gives

$$\begin{bmatrix} 1 & 2 \\ 3 & 4 \end{bmatrix}^2 - 5\begin{bmatrix} 1 & 2 \\ 3 & 4 \end{bmatrix} - 2\begin{bmatrix} 1 & 0 \\ 0 & 1 \end{bmatrix} = \begin{bmatrix} 0 & 0 \\ 0 & 0 \end{bmatrix}$$

Here's a simple program that illustrates the Cayley-Hamilton theorem for any $n \times n$ matrix.

```
def caley_hamilton(A):
    c = char_eqn(A)
    c = c.split('=')[0]                # left side of equation
    constant_term = ar.evaluate(c,'0','')[0]
            # subtract constant term:
    c = pa.main(c + '-(' + constant_term + ')')[0]
    c = c.replace('x','A')             # expression in A
            # add back constant term times I
    c = c + '+(' + constant_term + ')A^0'
    matrix_list = ['A='+A]    # ls.calculator requires this format
    return ls.calculator(c,matrix_list)

------------------------- Sample Run -------------------------
Input:
A = '1+3i,(2-7i)^3,3/7.12i; 4.8,5.9,6 -.8i;7/11,8.4,(9-.0987i)^2'
tl.format_print(caley_hamilton(A),2,'right')

Output:
0  0  0
0  0  0
0  0  0
```

12.11 Adjugate Matrix

The *adjugate* of a square matrix A, denoted by adj(A), is constructed as follows: First, for each i, j, calculate the determinant M_{ij} of the matrix obtained by removing row i and column j from A. Then multiply M_{ij}, the so-called *ij minor of A*, by $(-1)^{i+j}$ to obtain the i, j *cofactor A*. The matrix whose ij entry is C_{ij} is called the *cofactor matrix C* of A. For example,

$$A = \begin{bmatrix} 1 & 2 & 3 \\ 4 & 5 & 6 \\ 7 & 8 & 9 \end{bmatrix} \quad C = \begin{bmatrix} +\begin{vmatrix} 5 & 6 \\ 8 & 9 \end{vmatrix}, & -\begin{vmatrix} 4 & 6 \\ 7 & 9 \end{vmatrix}, & +\begin{vmatrix} 4 & 5 \\ 7 & 8 \end{vmatrix} \\ -\begin{vmatrix} 2 & 3 \\ 8 & 9 \end{vmatrix}, & +\begin{vmatrix} 4 & 6 \\ 7 & 9 \end{vmatrix}, & -\begin{vmatrix} 4 & 5 \\ 7 & 8 \end{vmatrix} \\ +\begin{vmatrix} 2 & 3 \\ 5 & 6 \end{vmatrix}, & -\begin{vmatrix} 1 & 3 \\ 4 & 6 \end{vmatrix}, & +\begin{vmatrix} 1 & 2 \\ 4 & 5 \end{vmatrix} \end{bmatrix}.$$

(We have inserted commas between entries so that we could illustrate the sign pattern without causing confusion.) The *adjugate* of A is defined as the transpose of the cofactor matrix of A.

The function adjugate(A) returns the adjugate of a matrix. The sample run illustrates the general property that the adjugate of A divided by the determinant d of A is the inverse of A (provided $d \neq 0$).

```
def adjugate(A):
    n = len(A)
    B = []
    for j in range(n):
        row = []
        for k in range(n):
            C = remove_row_col(A,j,k)
            d = det_echelon(C)
            sign = str((-1)**(j+k))
            d = ar.main('(' + sign + ')(' + d + ')')[0]
            row = row + [d]
        B.append(row)
    return ls.transpose(B)
```

```
-------------------------- Sample Run --------------------------
Input:
A = '2,2,3; \
     4,5,6; \
     7,8,9'
A = tl.string2table('2,2,3;4,5,6;7,8,9')
Adj = adjugate(A)
print('adjugate of A')
tl.format_print(Adj,2,'right'); print('\n')
B = ls.mult_mat(Adj,A)
print('A times the adjugate of A')
tl.format_print(B,2,'right'); print('\n')
print('determinant of A')
print(det_echelon(A))
-----------------------------------------------------------------
Output:
adjugate of A
-3   6  -3
 6  -3   0
-3  -2   2

A times the adjugate of A
```

12.11 Adjugate Matrix

```
-3   0   0
 0  -3   0
 0   0  -3

determinant of A
-3
```

The function `adjugate_params` is a version of `adjugate` that allows entries that are variables. It is gotten by replacing `ar.main` in the preceding function with `mu.main` and has the same defining property, as illustrated in the sample run using a modified version of `mult_mat`.

```
def adjugate_params(A):
    n = len(A)
    B = []
    for j in range(n):
        row = []
        for k in range(n):
            C = remove_row_col(A,j,k)
            d = det_params(C)
            sign = str((-1)**(j+k))
            d = mu.main('(' + sign + ')(' + d + ')')[0]
            row = row + [d]
        B.append(row)
    return ls.transpose(B)

def mult_mat_params(A,B):
    C = []
    for i in range(len(A)):                # run through rows of A
        Crow = []
        for j in range(len(B[0])):         # run through cols of B
            s = '0'
            for k in range(len(A[0])):
                s = s + '+('+A[i][k]+')('+B[k][j]+')'
            s = mu.main(s)[0]
            Crow.append(s)
        C.append(Crow)
    return C
```
```
--------------------------- Sample Run ---------------------------
Input:
A = '2,2x^2,3; \
     4x+y,5,6; \
     7,8,-3ax'
A = tl.string2table(A)
print('A:')
tl.format_print(A,2,'right'); print('\n')

Adj = adjugate_params(A)
```

```
print('adjugate of A:')
tl.format_print(Adj,2,'right'); print('\n')

B = mult_mat_params(Adj,A)
print('A times the adjugate of A:')
print(B); print('\n')
print('determinant of A:')
print(det_params(A))
```

```
Output:
A:
   2    2x^2    3
  4x+y   5      6
   7     8    -3ax

adjugate of A:
           -15ax-48    6ax^3+24              12x^2-15
     12ax^2+3axy+42    -6ax-21               12x+3y-12
         32x+8y-35    14x^2-16         -8x^3-2x^2y+10

A times the adjugate of A:
[['24ax^4+6ax^3y+84x^2-30ax+96x+24y-201', '0', '0'], \
 ['0', '6ax^3y+24ax^4+84x^2-30ax+24y+96x-201', '0'], \
 ['0', '0', '6ax^3y+24ax^4-30ax+84x^2+24y+96x-201']]

determinant of A:
6ax^3y+24ax^4-30ax+84x^2+96x+24y-201
```

The sample run shows that

$$\begin{bmatrix} 2 & 2x^2 & 3 \\ 4x+y & 5 & 6 \\ 7 & 8 & -3ax \end{bmatrix}^{-1}$$

$$= \frac{1}{6ax^3y + 24ax^4 + 84x^2 - 30ax + 24y + 96x - 201} \times$$

$$\begin{bmatrix} -15ax - 48 & 6ax^3 + 24 & 12x^2 - 15 \\ 12ax^2 + 3axy + 42 & -6ax - 21 & 12x + 3y - 12 \\ 32x + 8y - 35 & 14x^2 - 16 & -8x^3y - 2x2y + 10 \end{bmatrix}$$

Multivariable Algebra with Parameters

13

In this chapter we construct the module `MultiAlgParams.py`, which generalizes the module `MultiAlg.py` by allowing monomial coefficients to include parameters. Parameters, like scalars, are treated by the module as constants and placed in the first position of a monomial list. But unlike scalars, they have no given value during a particular program run. They are useful in several ways. First, they allow the user to test variations of a function to illuminate the nature of a particular process or model. Second, they are convenient in situations that require undetermined coefficients, for example the expansion of a rational function into a sum of partial fractions or the generation of integer summation formulas. The last two notions are explore in sections at the end of the chapter.

To avoid confusion, variables are required to be lower case letters and parameters upper case letters. For example

$$\frac{Ay}{xz-1} + \frac{Bx}{yz-2} + \frac{Cz}{xy-3}$$

is an expression of the required type, with A, B, C the parameters and x, y, z the variables.

The module `MultiAlgParams.py` is almost identical to `MultiAlg.py` but with obvious differences necessitated by the appearance of letters (parameters) in the coefficient entry of a monomial list. In the following sections we show how the relevant functions in `MultiAlg.py` must be modified. For convenience and to avoid possible conflicts we have included in the module `MultiAlgParams.py` functions from `MultiAlg.py` that require no change.

The module is headed by the import statements

```
------------------------- MultiAlgParams.py --------------------
import Number as nm
import Arithmetic as ar
import Tools as tl
import MultiAlg as mu
import LinSolve as ls
import PolyAlg as pa
import PolyDiv as pd
from operator import itemgetter
----------------------------------------------------------------
```

13.1 The Module

The following subsections describe the specific modifications needed for the inclusion of parameters.

Main Function

The function main(expr) takes the place of the eponymous function in MultiAlg.

```
----------------------------------------------------------------
def main(expr):
    global idx
    global varbs, varbs_list           # variable letters in expr
    global pars,pars_list              # parameter letters in expr
    varbs = tl.get_lower(expr)
    pars = tl.get_upper(expr)
    if varbs == ''  or   pars == '':
        return mu.main(expr)
    varbs_list = list(varbs)
    pars_list = list(pars)
    expr = tl.attach_missing_exp(expr,varbs+pars)
    expr = tl.fix_signs(expr)
    expr = tl.fix_operands(expr)
    expr = tl.insert_asterisks(expr,varbs+pars)
    idx = 0                                    # beginning of expr
    R = allocate_ops(expr,0)                   # make the calculations
    num = R[0]
    den = R[1]
    num = combine_monos(num)
    den = combine_monos(den)
    num = sort_list(num)
    den = sort_list(den)
    ratlist = [num,den]
    rat = list2rational(ratlist)
```

13.1 The Module

```
        irat, iratlist = list2int_rational(ratlist)
        return rat,ratlist,irat,iratlist

--------------------------- Sample Run ---------------------------
Input:
expr = 'Ay/(Bx-1.1) + Cx/(Dz-2.2)'
rat, ratlist, irat, iratlist = main(expr)
print(rat,'\n)
print(irat)                              # integer coefficients
------------------------------------------------------------------
Output:
(BCx^2+ADyz+(-11/10)Cx+(-11/5)Ay)/(BDxz+(-11/10)Dz+(-11/5)Bx+(121/50))
5(10BCx^2+10ADyz-11Cx-22Ay)/(50BDxz-55Dz-110Bx+121)
------------------------------------------------------------------
```

Calculations

As before, calculations are performed on multilists. Here are the revised calculation functions, obtained by substituting mu.main for ar.main to handle calculations involving parameters.

```
------------------------------------------------------------------
def combine_monos(P):
    Q = []
    for i in range(len(P)-1):
        if P[i] == '': continue            # already added
        M = P[i]                           # ith monomial: [coeff,powers]
        if M == [] or  M[0] == '0':
            continue
        for j in range(i+1,len(P)):  # add succeeding like monos to M
            if P[j] == '': continue
            N = P[j]
            if M[1:] == N[1:]:   # if powers the same, add the coeffs
                coeffsum = mu.main(M[0] + '+(' + N[0] +')')[0]
                M[0] = coeffsum                # update M's coefficient
                P[j] = ''                      # mark as already added
        Q.append(M)   # append nonzero monomial in P[i]
    leftover_mono = P[len(P)-1]
    if leftover_mono != '' and leftover_mono[0] != '0':
        Q.append(leftover_mono)   # pick up leftover monomial at end
    return Q
------------------------------------------------------------------
```

The change in the allocator function is the inclusion of the following elif statement, which converts a parameter into a rational list.

```
        elif ch in pars:
            idx +=1
            exp,idx = tl.extract_exp(expr, idx)
            exp = ar.main(exp)[0]
            ch = ch + '^' + exp
            R = scalar2rat(ch,len(varbs))
```

List to Monomial

The difference here is the use of mu.main:

```
def list2monomial(Mlist):
    # takes monomial list and returns monomial
    expr = ''
    coeff = Mlist[0]
    coeff = mu.main(coeff)[0]
    if coeff == '0' or coeff == '-0':
        return '0'
    if len(Mlist) == 1:
        return coeff
    mono = ''
    for i in range(1,len(Mlist)):
        var  = varbs_list[i-1]
        exp = Mlist[i]                       # exponent of variable
        #print(2224,i,var,exp)
        if exp > 1:
            mono = mono + var + '^' + str(exp)    # attach exp != 1
        elif exp == 1:
            mono = mono + var
    if coeff == '1':
        coeff = ''                           # coeff '1' not needed
    if coeff == '-1':
        coeff = '-'
    if coeff != '' or exp != 0:
        coeff = tl.add_parens(coeff)
    expr = coeff +  mono
    expr = tl.fix_signs(expr)
    return expr
```

Integer Coefficients

Here is the replacement for the function that converts to integer coefficients.

```
def list2int_rational(R):
    numR = R[0]; denR = R[1]
    if numR == denR:
        return '1',''
    lcmP,P = clear_pol(numR)
    lcmQ,Q = clear_pol(denR)
    factor = mu.main(lcmQ +'/'+ lcmP)[0]
    irat_list = [P,Q]
    irat = list2rational(irat_list)
    if factor == '1': factor = ''
    factor = tl.add_parens(factor)
    irat = tl.add_parens(irat)
    return factor+irat, irat_list
```

Evaluating an Expression

The evaluation function takes the following form in the parameter setting.

```
def evaluate(expr,substitutions):
    substitutions = substitutions.split(',')
    for i in range(len(substitutions)):
        substitutions[i] = substitutions[i].replace(' ','')
        var,val = substitutions[i].split('=')
        if var == '' or var not in expr: continue
        expr = expr.replace(var,'(' + val + ')')
    expr = tl.fix_signs(expr)
    if tl.has_no_letters(expr):
        return ar.main(expr)[0],'','',''
    return mu.main(expr)
```

13.2 Application: Sums of Integer Powers

In this section we construct a program that finds a closed expression for sums of the form

$$S(n, p) = 1^p + 2^p + \cdots + n^p$$

where n and p are positive integers. To see how this works, consider the case $p = 2$. We assume that $S(n, 2)$ is a polynomial in n of degree 3 with no constant term (a leap of faith):

$$S(n, 2) = An^3 + Bn^2 + Cn$$

We then have
$$S(n+1, 2) = A(n+1)^3 + B(n+1)^2 + C(n+1).$$

On the other hand,
$$S(n+1, 2) = S(n, 2) + (n+1)^2 = (An^3 + Bn^2 + Cn) + (n+1)^2.$$

Subtracting the two versions of $S(n, 2)$ and collecting like powers yields

$$\begin{aligned} 0 &= A\big((n+1)^3 - n^3\big) + B\big((n+1)^2 - n^2\big) + C\big((n+1) - n\big) - (n+1)^2 \\ &= A(3n^2 + 3n + 1) + B(2n + 1) + C - (n^2 + 2n + 1) \\ &= (3A - 1)n^2 + (3A + 2B - 2)n + (A + B + C - 1). \end{aligned} \quad (13.1)$$

Equating coefficients to zero we obtain the system

$$\begin{aligned} 3A &= 1 \\ 3A + 2B &= 2 \\ A + B + C &= 1 \end{aligned} \quad (13.2)$$

The solutions are easily seen to be $A = 1/3$, $B = 1/2$, and $C = 1/6$, leading to the closed formula
$$S(n, 2) = n^3/3 + n^2/3 + n/6 = n(n+1)(2n+1)/6.$$

Several functions are needed to implement the algorithm. The first forms the differences, as in (13.1), but in list form.

```
---------------------------------------------------------------
def pol_diff(p):
    term = ''
    for k in range(p+1):
        letter = t1.upper[k]
        power = p-k+1
        term = term +'+'+ letter +'((n+1)^'+ str(power) +'-n^'\
            + str(power) +')'
    term = term + '- (n+1)^' + str(p)
    return main(term)[1][0]                        # no denominator
--------------------------- Sample Run ------------------------
Input:
print(pol_diff(2))
---------------------------------------------------------------
Output:
[['3A-1', 2], ['3A+2B-2', 1], ['A+B+C-1', 0]]
---------------------------------------------------------------
```

13.2 Application: Sums of Integer Powers

The next function equates to zero the coefficient expressions in the output of `pol_diff(p)` and then uses `linsolve` to find the coefficients.

```
def get_coefficients(p):
    eqns = ''
    pol_diff_list = pol_diff(p)
    for k in range(p+1):
        eqns += pol_diff_list[k][0] + '= 0,'
    coeffs = ls.linsolve(eqns[:len(eqns)-1],'','',False)
    return coeffs
```
---------------------------- Sample Run ----------------------------
Input:
`print(get_coefficients(2))`

Output:
`['A=1/3', 'B=1/2', 'C=1/6']`

The final function creates the formula from the coefficient. Print statements chronical the evolution of the formula.

```
def make_formula(p):
    coeffs = get_coefficients(p)    # e.g. ['A=1/3','B=1/2','C=1/6']
    formula = ''
    L = len(coeffs)
    for k in range(L):
        if coeffs[k] == '0': continue
        coeffs[k] = coeffs[k].split('=')[1]    # e.g. 1/3,1/2,1/6
        coeffs[k] = tl.add_parens(coeffs[k])
        formula = formula + '+' + coeffs[k] + 'n^' + str(L-k)
    formula = formula.replace('^1','')
    formula = tl.fix_signs(formula)
    formula = pa.main(formula)[4]    # integer coefficient pol
    formula = pd.factor_polynomial(formula)
    return formula
```
---------------------------- Sample Run ----------------------------
Input:
`print(make_formula(11))`

Output:
`((1/12)(n^8+4n^7+2n^6-8n^5+(-5/2)n^4+13n^3+(-3/2)n^2-10n+5))(n+1)^2n^2`

13.3 Application: Partial Fractions

A *partial fraction expansion* of a rational function $R(x) = P(x)/Q(x)$ is a sum of a polynomial $P_0(x)$ and terms $R_k(x)$ that are rational functions with denominator degree at most 2 and numerator degree at most 1:

$$R(x) = P_0(x) + R_1(x) + R_2(x) + \cdots + R_m(x).$$

The expansion finds its most important use in finding the integral of a rational function. If $\deg P \geq \deg Q$, then $P_0(x)$ may be found by the division algorithm. Accordingly, we consider only the case $\deg P < \deg Q$.

For the method to work it is necessary that $Q(x)$ be factored into a product of linear factors $ax + b$ and irreducible quadratic factors $ax^2 + bx + c$. The term *irreducible* means that the equation $ax^2 + bx + c = 0$ has no real solutions, that is, $b^2 - 4ac < 0$. In the discussion that follows we consider only the case where $a = 1$, since this may be achieved by factoring out the a's in Q and adjusting $P(x)/Q(x)$ accordingly, for example,

$$\frac{P(x)}{(3x+2)(5x^2+3x+1)} = \frac{1}{15} \frac{P(x)}{(x+2/3)(x^2+3/5x+1/5)}.$$

The following examples demonstrate the various cases that occur in partial fraction expansions. In each case there are unique constants $A, B, \ldots$ for which the equation holds.

Case 1: Q is a product of distinct linear factors:

$$\frac{2x+5}{(x-1)(x+2)} = \frac{A}{x-1} + \frac{B}{x+2}.$$

Case 2: Q is a product of linear factors, some repeated:

$$\frac{2x+5}{(x-1)^2(x+2)} = \frac{A}{x-1} + \frac{B}{(x-1)^2} + \frac{C}{x+2}.$$

Case 3: Q is a product of distinct irreducible quadratic factors:

$$\frac{x^3+2}{(x^2+x+1)(x^2+1)} = \frac{Ax+B}{x^2+x+1} + \frac{Cx+D}{x^2+1}.$$

Case 4: Q is a product of quadratic factors, some repeated:

$$\frac{x^3+5}{(x^2+x+1)(x^2+1)^2} = \frac{Ax+B}{x^2+x+1} + \frac{Cx+D}{x^2+1} + \frac{Ex+F}{(x^2+1)^2}.$$

Case 5: Q is a mix of linear and irreducible quadratic factors:

$$\frac{7x^2+3}{(x+1)^2(x^2+2x+5)} = \frac{A}{(x+1)} + \frac{B}{(x+1)^2} + \frac{Cx+D}{x^2+2x+5}.$$

13.3 Application: Partial Fractions

To find the values of the constants $A, B, \ldots$, one clears fractions, subtracts the left side from the right, collects together coefficients of like powers, sets each of the collected coefficient expressions to zero, and solves the resulting linear system.

We illustrate with the example in Case 5. Multiplying the equation by the denominator of the left side yields

$$7x^2 + 3 = A(x+1)(x^2 + 2x + 5) + B(x^2 + 2x + 5) + (Cx + D)(x+1)^2,$$

which we write as

$$A(x+1)(x^2 + 2x + 5) + B(x^2 + 2x + 5) + (Cx + D)(x+1)^2 - (7x^2 + 3) = 0.$$

Expanding the left side and collecting coefficients of like powers we have

$$(A + C)x^3 + (3A + B + 2C + D - 7)x^2 + (7A + 2B + C)x + (5A + 5B + 2D - 3) = 0.$$

Setting the coefficients to zero we obtain the system

$$A + C = 0$$
$$3A + B + 2C + D - 7 = 0$$
$$7A + 2B + C = 0$$
$$5A + 5B + 2D - 3 = 0$$

The solutions are then plugged into the equation in Case 5.

To implement the procedure in Python the coefficients of the rational function need to be Gaussian rationals. The user enters the rational function $R = P/Q$, with Q factored as explained above; the program returns its partial fraction expansion. The main function, `partial_fractions`, carries out the general technique indicated in the above example. For clarity, the code includes print statements to illustrate the steps.

```
def partial_fractions():
    global equations, partials_cleared

    print('Step 1: get numerator and denominator of rational:')
    get_numerator_denominator()
    print(numerator,',',denominator,'\n')

    print('Step 2: get denominator factors:')
    get_denominator_factors()
    print(denominator_factors,'\n')

    print('Step 3: make template:')
    make_template()
    print(template,'\n')
```

```
            print('Step 4: clear partial fractions:')
            clear_partial_fractions()
            tl.print_list(partials_cleared,'v')
            print('\n')

            print('Step 5: make expression from cleared partial fractions:')
            make_expression()
            print(expr,'\n')

            print('Step 6: make equations:')
            make_equations()
            tl.print_list(equations,'v')
            print('\n')

            print('Step 7: get letter values:')
            get_letter_values()
            print(letter_values,'\n')

            print('Step 8: print the expansion:\n')
            print_expansion()
```

Here is a sample run.

```
    Input:
    rational = '(7x^2+3)/(x+1)(x+2)(x^2+2x+5)^3'
    partial_fractions()

    Output:
    Step 1: get numerator and denominator of rational:
    (7x^2+3), (x+1)(x+2)(x^2+2x+5)^3

    Step 2: get denominator factors:
    ['(x+1)', '(x+2)', '(x^2+2x+5)^3']

    Step 3: make template:
    [['A', '(x+1)'], ['B', '(x+2)'], ['(Cx+D)', '(x^2+2x+5)'],
    ['(Ex+F)', '(x^2+2x+5)^2'], ['(Gx+H)', '(x^2+2x+5)^3']]

    Step 4: clear partial fractions:
    x^7+8x^6+39x^5+122x^4+271x^3+420x^2+425x+250
    x^7+7x^6+33x^5+95x^4+203x^3+285x^2+275x+125
    x^6+7x^5+28x^4+70x^3+113x^2+115x+50
    x^4+5x^3+13x^2+19x+10
    x^2+3x+2

    Step 5: make expression from cleared partial fractions:
    A(x^7+8x^6+39x^5+122x^4+271x^3+420x^2+425x+250)+
    B(x^7+7x^6+33x^5+95x^4+203x^3+285x^2+275x+125)+
```

13.3 Application: Partial Fractions

```
(Cx+D)(x^6+7x^5+28x^4+70x^3+113x^2+115x+50)+
(Ex+F)(x^4+5x^3+13x^2+19x+10)+(Gx+H)(x^2+3x+2)

Step 6: make equations:
A+B+C= 0
8A+7B+D+7C= 0
39A+33B+28C+7D+E= 0
122A+95B+70C+28D+F+5E= 0
271A+203B+113C+70D+13E+5F+G= 0
420A+285B+115C+113D+19E+13F+3G+H-7= 0
425A+275B+50C+115D+10E+19F+3H+2G= 0
250A+125B+50D+10F+2H-3= 0

Step 7: get letter values:
[['A', '5/32'], ['B', '-31/125'], ['C', '367/4000'],
 ['D', '-5/32'], ['E', '123/200'], ['F', '-5/8'],
 ['G', '37/10'], ['H', '9/2']]

Step 8: print the expansion:
        (7x^2+3)
  ---------------------
  (x+1)(x+2)(x^2+2x+5)^3

      5/32
  = -----
     (x+1)


     -31/125
  + -------
      (x+2)

     (367/4000)x-5/32
  + ----------------
       (x^2+2x+5)

     (123/200)x-5/8
  + --------------
      (x^2+2x+5)^2

     (37/10)x+9/2
  + ------------
     (x^2+2x+5)^3
```

For the remainder of the section we describe the functions used in steps 1–8. The first, get_numerator_denominator, takes the original rational expression and splits it into numerator and denominator.

```
def get_numerator_denominator():
    global numerator, denominator, rational
    rational = rational.replace(' ','')
    numerator, denominator = rational.split('/')
```

The function get_denominator_factors splits the factors of the denominator of the rational into a list. It does so by inserting a comma before each left parenthesis and then splits the denominator at the newly installed comma.

```
def get_denominator_factors():
    global denominator, denominator_factors
    denominator_factors = denominator.replace('(',',(')
    denominator_factors = denominator_factors.split(',')
    denominator_factors = denominator_factors[1:] #avoid 1st comma
```

The function make_template creates a list which is used later to form the partial fraction expansion. For example, the function takes the factor (x²+x+1) and produces the list ['Ax+B','(x²+x+1)']

```
def make_template():
    global template, var
    template= []
    var = tl.get_var(rational)
    k = 65                                       # ASCII code for 'A'
    for den in denominator_factors:
        if ')^' not in den:                      # for uniformity
            den = den+'^1'                       # e.g. (x^2+x+1) --> (x^2+x+1)^1
        factor,exp = den.split(')^')             # (x^2+x+1)^2 --> (x^2+x+1, 2
        factor = factor + ')'                    # --> (x^2+x+1)
        if '^2' not in factor:                   # linear factor
            for i in range(1, int(exp)+1):
                if i == 1:                       # (x+1) -->A(x+1)
                    template.append([chr(k), factor])
                else:                            # (x+1) -->A(x+1)^i
                    template.append([chr(k), factor +'^'+ str(i)])
                k += 1
        else:                                    # quadratic factor
            for i in range(1, int(exp)+1):
                if i == 1: # (x^2 + x + 1) --> Ax+B, (x^2 + x + 1)
                    template.append(['('+ chr(k) + var +'+'+     \
                                             chr(k+1) +')',factor])
                else:     # (x^2 + x + 1) --> Ax+B, (x^2 + x + 1)^i
```

13.3 Application: Partial Fractions

```
                template.append([' ('+ chr(k) + var +'+'+ \
                                 chr(k+1) +')', \
                                 factor +'^'+ str(i)])
        k += 2
    if k == 90:                                    # ran out of caps
        break
```

The function `clear_partial_fractions` effectively takes the right side of the expansion and multiplies it by the denominator of the given rational expression. For example, for the rational function $(7x^2 + 3)/(x + 1)^2(x^2 + 2x + 5)$, which has the expansion

$$\frac{A}{(x+1)} + \frac{B}{(x+1)^2} + \frac{Cx+D}{(x^2+2x+5)},$$

the template is

```
[['A', '(x+1)'], ['B', '(x+1)^2'], ['(Cx+D)', '(x^2+2x+5)']]
```

The function calculates

$$\frac{(x+1)^2(x^2+2x+5)}{(x+1)} = x^3 + 3x^2 + 7x + 5$$

$$\frac{(x+1)^2(x^2+2x+5)}{(x+1)^2} = x^2 + 2x + 5$$

$$\frac{(x+1)^2(x^2+2x+5)}{(x^2+2x+5)} = x^2 + 2x + 1$$

and returns the right sides of these equations in the list `partials_cleared`.

```
def clear_partial_fractions():
    global partials_cleared
    partials_cleared = []
    for t in template:             # e.g. t = ['(Cx+D)', '(x^2+2x+5)']
        den = t[1]                           # '(x^2+2x+5)
        Q,R = pd.div_alg(denominator,den)
        partials_cleared.append(Q)
```

The following function takes the list `partials_cleared` and attaches the letter numerators from the template.

```
def make_expression():
    global expr
    expr = ''
    for i in range(len(partials_cleared)):
        prod = template[i][0] + '('+ partials_cleared[i] +')'
        expr = expr + '+' + prod
    expr = expr[1:]                              # remove first plus
```

The function make_equations uses main to expand diff=expr-numerator into standard polynomial form and then equates the coefficients to zero.

```
def make_equations():
    global equations
    equations = ''
    make_expression()
    diff = expr +'-'+ numerator
    diff = main(diff)[1][0]
    for entry in diff:
        eqn = entry[0] + '= 0'
        equations = equations + ',' + eqn
    equations = equations[1:]                    # remove initial comma
```

The function get_letter_values feeds the equations formed by the preceding function to ls.linsolve, which provides values for the letters in the partial fraction expansion.

```
def get_letter_values():
    global letter_values
    letter_values = []
    letter_vals = ls.linsolve(equations,'','')[0]
    for item in letter_vals:
        letter,value = item.split('=')
        letter = letter.replace(' ','')
        value = value.replace(' ','')
        letter_values.append([letter,value])
```

13.3 Application: Partial Fractions

The function `substitute_values` takes the values obtained in the preceding function and inserts them into the template letter entries.

```
def substitute_values():
    for j in range(len(template)):
        num = template[j][0]
        for item in letter_values:
            if item[0] in num:
                item[1] = tl.add_parens(item[1])
                template[j][0] = \
                    template[j][0].replace(item[0],item[1])
```

The function `print_expansion` prints the final partial fraction expansion.

```
def print_expansion():
    global template #letter_values
    substitute_values()
    tl.print_fraction('',numerator,denominator,'')        # rational
    print('\n')
    for j in range(len(template)):
        num = template[j][0]
        den = template[j][1]
        num = pa.polycalc(num)[0]                          # clean up
        if num == '0':                                     # nothing to print
            continue
        if j == 0:
            tl.print_fraction(' = ',num,den,'')            # print 1st term
            print('\n')
        else:
            tl.print_fraction(' + ',num,den,'')            # print the rest
            print('\n')
```

Index

A
Absolute value, 5
Argument, 1, 49
ASCII, 10

B
Base case, 22
Basis, 207
Bézout coefficients, 62
Bool type, 2

C
Cayley-Hamilton theorem, 226
Characteristic equation, 226
Characteristic polynomial, 226
Chr function, 11
Cipher, 92
Clearing the column, 160
Coefficient, 95
Coefficient matrix, 187
Comparison operators, 6
Complex type, 2
Components of a vector, 201
Composite, 60
Compound statement, 41
Concatenation operator, 7
Conclusion, 49
Conditional statement, 6

Congruent, 70
Constant term, 95
Contradiction, 48
Co-planar, 222
Count method, 9

D
Degree, 95, 132
Dependency relation, 203
Dimension, 207
Dividend, 5, 120
Division algorithm, 5
Divisor, 5, 119, 120

E
Eigenspace, 226
Eigenvalue, 226
Equation operations, 157
Equivalent stmts, 49
Eratosthenes of Cyrene, 68
Euclidean distance, 191

F
Factors, 119
Find method, 10
Float function, 3
Float type, 2

Floor division, 5
Function, 1

G
Gaussian rational number, 75
Greatest common divisor, 60

H
Hierarchy, 42

I
Identity matrix, 182
Immutable, 10
Index, 8
Index method, 12
Interpolation function, 106
Int function, 3
Int type, 2
Irreducible, 238

J
Join method, 13

K
Keyword, 3

L
Leading coefficient, 95
Least squares method, 193
Least squares solution, 190
Length function, 8
Linear combination, 202
Linearly dependent, 203
Linearly independent, 203
List function, 13, 16
Logical operators, 6
Lower method, 9

M
Matrix, 156
Matrix dimensions, 156
Membership operator, 7

Method, 2
Modulo operator, 5
Monic polynomial, 95
Monomial, 95, 131
Multivariate polynomial, 132

N
n-dimensional vector, 201

O
Ord function, 11

P
Parameter, 1
Pivot column, 160
Pivot position, 160
Pivot row, 160
Polynomial of degree m, 95
Precedence, 42
Precedence rules, 4
Premise, 49
Prime, 60
Print function, 2
Proposition, 41

Q
Quotient, 5, 120

R
Range statement, 21
Reduced row echelon form, 159
Remainder, 5, 120
Replace method, 9
Root, 123
Row operations, 158

S
Scientific notation, 76
Script, 1
Set function, 16
Sign of a permutation, 212
Slice function, 8, 12
Span, 207

Split method, 13
Statement, 41
Str function, 3
Str type, 2
Substring, 8

T
Table, 34
Tautology, 48
Taylor series expansion, 110
Term, 95
Terminating condition, 22
Transpose, 166
Truth table, 42
Type, 2
Type function, 3

U
Unique prime factorization theorem, 68
Upper method, 9
Upper triangular matrix, 214

V
Valid argument, 49
Vandermonde determinant, 195
Vandermonde equation, 195
Vandermonde matrix, 195
Variable, 3
Vector space, 207

Z
Zero, 123
Zero vector, 201

GPSR Compliance

The European Union's (EU) General Product Safety Regulation (GPSR) is a set of rules that requires consumer products to be safe and our obligations to ensure this.

If you have any concerns about our products, you can contact us on ProductSafety@springernature.com

In case Publisher is established outside the EU, the EU authorized representative is:

Springer Nature Customer Service Center GmbH
Europaplatz 3
69115 Heidelberg, Germany

Batch number: 08759005

Printed by Printforce, the Netherlands